THE NEW SIMPLE AND PRACTICAL SOLAR COMPONENT GUIDE

by Lacho Pop, MSE and
Dimi Avram, MSE

Digital Publishing

Disclaimer Notice

The authors of this ebook, named '**The New Simple and Practical Solar Component Guide**', hereinafter referred to as the 'Book', make no representation or warranties with respect to the accuracy, applicability, fitness, or completeness of the contents of the Book. The information contained in the Book is strictly for educational purposes. Summaries, strategies, tips and tricks are only recommendations by the authors, and the reading of the Book does not guarantee that reader's results shall exactly match the authors' results.

The authors of the Book have made all reasonable efforts to provide current and accurate information for the readers of the Book, and the authors shall not be held liable for any unintentional errors or omissions that may be found.

The Book is not intended to replace or substitute any advice from a qualified technician, solar installer or any other professional and advisor, nor should it be construed, as legal or professional advice and the authors explicitly disclaim any responsibilities for such use.

The installation of solar power systems requires certain background professional qualification and certification for working with high voltages and currents dangerous to human life and for installing solar power systems and appliances. The reader should consult every step of your project or installation with a qualified solar professional, installer or technician and local authorities.

The authors shall in no event be held liable to any

About the Authors

Lacho Pop, MSE, has more than 15 years of experience in market research, technological research and design and implementation of various sophisticated electronic and telecommunication systems. His large experience helps him present the complex world of solar energy in a manner that is both practical and easy understood by a broad audience.

Dimi Avram, MSE, has more than 10 years of experience in engineering of electrical and electronic equipment. He has specialized in testing electronic equipment and performing a techno-economic evaluation of various kinds of electric systems. His excellent presentation skills help him explain even the most complicated stuff to anybody interested.

Table of Contents

Introduction

Have you ever wanted to save money on electricity and become energy-independent?
Do you want to protect your family from regular power outages and negligence of local utility?

The book "The New Simple and Practical Solar Component Guide" helps you accomplish this by understanding the essentials of building blocks of solar systems and harnessing solar power in your needs.

Written by electronic engineers, this easy-to-read-and-follow solar components guide demystifies all of the components of a solar power system in a way that anyone without a technical background can understand. The book is useful for a broad audience: technically and non-technically inclined people, beginners and advanced in solar power, and electrical engineers.

Based on thousand hours of research and experience, the book contains practical solar power information you cannot find and cannot apply just by searching the web.

This book is focused on:

- o electricity basics related to solar power
- o solar power system types
- o solar power system components:
  - • solar panel module types
  - • bypass diodes
  - • blocking diodes
  - • inverters
  - • microinvertes

1

- charge controllers
- batteries
- fuses
- surge protectors/arrestors
- disconnects (circuit breakers)
- junction (combiner) boxes
  - cables

   ... and more.

Furthermore, this guide provides you with a serious introduction and practical information to solar power, coupled with action-oriented how-to tips and guidelines about selecting, combining and sizing solar components.

This book, however, is not a complete A-Z guide for sizing a solar power system. Every journey starts with most important first steps, and having a sound basis about solar components is a prerequisite for getting an efficient safe and cost effective solar power system for your home, RV vehicle, boat or business. Without this essential knowledge, every attempt for designing and sizing a solar power system will be either fruitless or counterproductive.

Photovoltaics: How they work

A solar photovoltaic (PV) system converts sunlight into electricity (electrical energy).

Here terms 'solar' and 'photovoltaic' are used interchangeably. However 'solar' is also used for solar water-heating systems which are beyond the scope of this book:

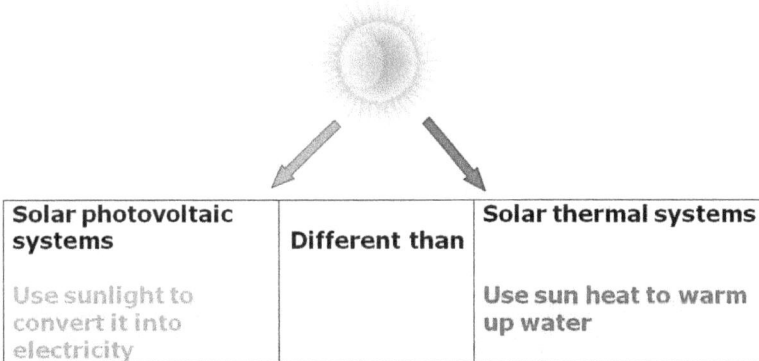

Solar photovoltaic systems	Different than	Solar thermal systems
Use sunlight to convert it into electricity		Use sun heat to warm up water

A photovoltaic system can generate a part of or all of your electricity demand.

It can either reduce the amount of power you consume from your utility or replace the utility grid completely, depending on how far you live from a utility grid.

The main component of every photovoltaic system is a photovoltaic (solar) module. A solar module consists of solar cells connected together. Furthermore, to achieve a higher energy yield solar modules can be connected in arrays.

In the picture below, you can distinguish between the main types of photovoltaic units.

PV cell

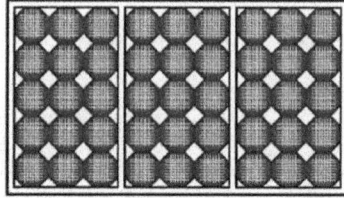

PV panel

PV module

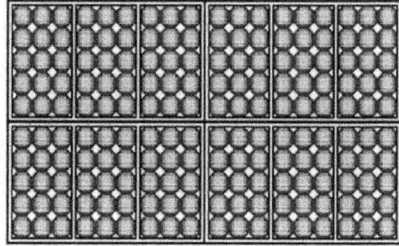

PV array (PV generator)

**PV Cell -> PV Module -> PV Panel ->
PV Array -> PV System**

Photovoltaic cells and modules are made of semiconductor (usually silicon) capable of producing electricity when exposed to sunlight:

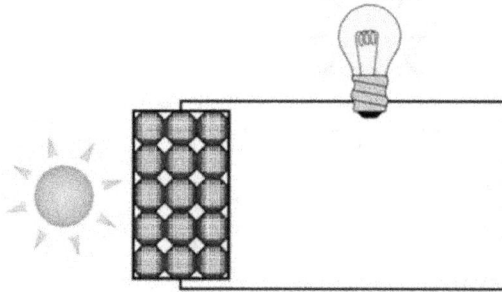

Such a capability is called 'photovoltaic effect.'

More energy is generated in sunny days while less energy is generated in cloudy/rainy days, or when the photovoltaic array is shaded by obstructions (trees, lampposts, buildings, etc.).

4

Some types of solar cells and modules, however, are capable of providing more electricity in cloudy days, since they react more heavily to ultraviolet component of the sunshine spectrum which is higher on such days.

Important!

PV modules generate DC electricity. To learn more about DC electricity refer to chapter 'Electricity basics').

The generated electrical energy can:

- Used right away
- Stored in a battery for later use
- Converted to AC electricity and then either used by home appliances or exported to the utility grid.

Electricity Basics

Voltage, current, power, and energy

There are two types of electrical devices – loads and generators.

Every photovoltaic module is a generator. It generates voltage when sunlight falls onto its surface.

Voltage is what makes electrical devices operate.

In a closed circuit with voltage applied, **current** starts flowing, and the plugged-in devices start to work.

Voltage and current are the main electrical parameters. Voltage is measured in volts (V) while current is measured in amps (A).

Power is the rate of generating or consuming energy. It is measured in Watts.

If your TV set is 150 W, it consumes 150 W at a given moment. Your TV set is a load – it consumes electricity.

If you have a photovoltaic array of 1,000 W installed on your roof, your PV system can generate 1,000 W at a given moment. Your PV system is a generator – it produces (generates) electricity.

Energy is the amount of work that can be done. Energy is measured in Watt-hours (Wh). Often kWh (equal to 1,000 Wh), rather than Wh, is used. To compare, 1 kWh equals 1,000 Wh.

Energy is the amount of electricity used or produced over a certain period.

If you know the rated power of that device and its

time of use, then you can calculate the energy consumed or generated by a device.

Example:

A TV set of 150 W, operating for 2 hours, will consume 150*2 = 300 Wh of energy.

A photovoltaic array of 1,000 W installed power, operating for 10 hours, will generate 1,000*10 = 10,000 Wh of energy.

When you get your monthly electricity bill, you pay for the total energy consumed by your household or office.

Such total consumed energy is obtained by multiplying the rating of every device by the number of hours it had been on during a month.

Example:

Here is an example list of daily loads:

Device	Rated power	Period of use	Energy consumed
TV set	100 W	3 h	300 Wh
Laptop	60 W	8 h	480 Wh
Lights	15 W	12 h	180 Wh
Radio	40 W	10 h	400 Wh
Vacuum cleaner	700 W	0.5 h	350 Wh
		Total energy consumed:	1,710 Wh

If this is the regular everyday consumption, then every month the user has to pay for 1,710 Wh * 30 days = 51,300 Wh = 51.3 kWh of energy consumed.

Typical power consumption of some standard household appliances

Source: http://www.wholesalesolar.com

Appliance	Wattage	Appliance	Wattage
Laptop	60-250	Hedge trimmer	450
LCD TV	213	1/2" drill	750
Plasma TV	339	1/4" drill	250
12" black and white TV	20	1" drill	1000
19" color TV	70	12" chain saw	1100
25" color TV	150	14" band saw	1100
Satellite dish	30	Waterpik	100
Cell Phone - recharge	2-4 watts	Weed eater	500
Computer	120	Well Pump (1/3-1 HP)	480-1200
Monitor	150	3" belt sander	1000
Standard TV	188	7-1/4" circular saw	900
Stereo	60	8-1/4" circular saw	1400
Video Game Player	195		
MP3 Player - recharge	0.25-0.40 watts	**Heaters**	
Radiotelephone - Receive	5	Electric Clothes Dryer	3,4
Radiotelephone - Transmit	40-150	Engine Block Heater	150-1000
Clock Radio	7	Stock Tank Heater	100
Blender	300	Furnace Blower	300-1000
Cable Box	20	Hot Plate	1200
Can Opener	100	Iron	1,1
Ceiling Fan	100	Portable Heater	1500
Central Air Conditioner	5	Toaster	1,1
Coffee Machine	1,5	Toaster oven	1,2
Curling Iron	90	Water heater	479
Dehumidifier	350	Room Air Conditioner	1,1
Humidifier	300-1000		
Dishwasher	1200-1500	**Lights**	
Electric blanket	200	100 watt incandescent bulb	100
Espresso Machine	360	20 watt DC compact fluor.	22
Hair Dryer	1,538	25 watt compact fluor. bulb	28
Shaver	15	CFL Bulb (60-watt equivalent)	18
Vacuum Cleaner	500	CFL Bulb (75-watt equivalent)	20
Microwave	1,5	40 watt DC halogen	40
		50 watt DC incandescent	50
Refrigerator/ Freezer			
16 cu. ft. (AC)	1200 Wh/day		
20 cu. ft. (AC)	1411 Wh/day		
Freezer			
15 cu. ft. (Chest)	1080 Wh/day		
15 cu. ft. (Upright)	1240 Wh/day		

What is important when buying an electrical device?

Important!

When buying an electrical device, you should be careful about:

- The voltage type (AC or DC)
- The rated voltage (120V or 240V for AC; 12V, 24V or 48V for DC).

Otherwise, the appliance either will get damaged or will not work!

DC and AC electricity

There are two types of electricity – direct (DC) and alternating (AC).

A device cannot operate on both AC and DC simultaneously – it only operates on one of them.

However, DC electricity can be converted into AC electricity by a device called an 'inverter.' An AC device cannot operate on DC electricity, and DC device cannot operate on AC electricity.

You have alternating (AC) electricity at your home and office.

It is called 'alternating,' since current and voltage change their direction over time:

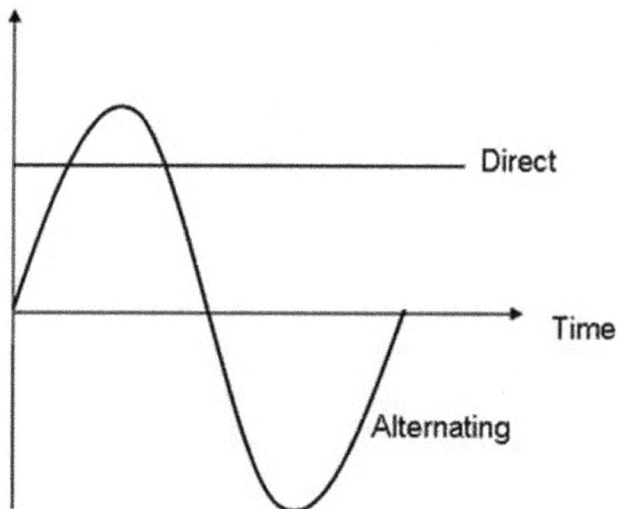

AC electricity:
- Is provided by public utility companies
- Is produced by fuel generators
- Cannot be stored for later use

What to remember about DC electricity

- Current and voltage have a constant direction
- Produced by batteries
- Can be stored for later use
- Used by certain devices (radios, radio clocks, flashlights, water pumps, DC fans, electrical shavers, etc.)

Important!

PV modules produce DC electricity that:

- Can be consumed right away by DC loads
- Can be stored in batteries for later use
- Can be converted into AC electricity and then either used by AC devices or exported to utility grid.

Electrical circuits and voltage drops

An electric circuit is a fixed, closed path where electric current flows from a voltage source (battery, PV module or AC generator) through a conductor (cable or wire) and a load.

When you plug the battery into a circuit, by adding load and connecting cables, you close the circuit. Upon the circuit being closed and the voltage **V** applied, the current **I** starts to flow:

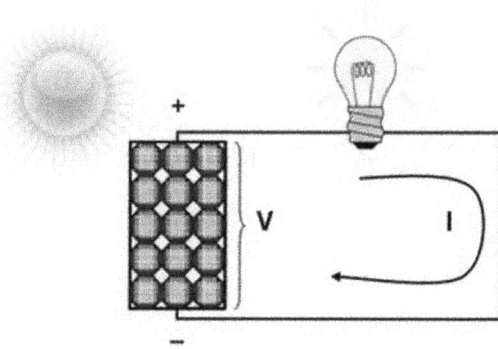

You get a closed circuit when you connect the voltage source to a load with a piece of wiring.

Each section in a closed circuit creates a **voltage drop**.

- A voltage drop has a value always lower than the battery voltage.
- The source voltage is the sum of all voltage drops created across a circuit.

Each component of the circuit, where current flows through, creates a voltage drop.

A load also creates a voltage drop. The voltage drop on a load is the voltage needed by the load to operate properly.

Unfortunately, voltage drops are also formed on wiring, which is undesired. Voltage drops on wiring are called 'voltage losses.'

The higher the voltage drop created on the wiring, the lower the voltage drop created on the load.

Important!

If the voltage drop on the wires is too high, the voltage drop on the load might turn out to be insufficient for that load to operate normally.

In such a situation, there are two options available:

- Choose a battery of a higher voltage to allow a higher voltage to drop on wires and sufficient operating voltage for the load.
- Use wires that form a lower voltage drop to deliver sufficient voltage to the load.

The sum of the voltage drops in a circuit is constant. Such a sum is the voltage of the power source.

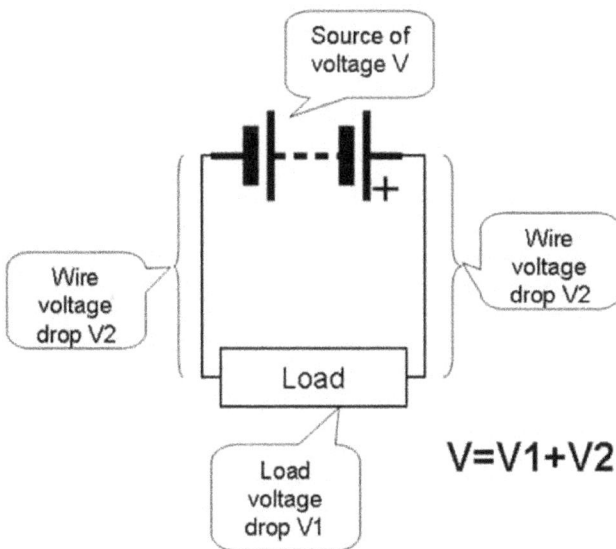

Source of voltage V

Wire voltage drop V2

Wire voltage drop V2

Load

Load voltage drop V1

$V=V1+V2$

Let's see an example.

If you use longer wires, they will produce a higher voltage drop V_2. Since the voltage V of the source remains the same, if V_2 increases (according to the formula $V=V_1+V_2$), the voltage drop V_1 on the load decreases.

So, from a certain moment on, the load might not work at all, since it cannot get the voltage it needs!

What does all this mean? When you add a new voltage drop in a closed circuit, all the existing voltage drops decrease, since the sum of all is supposed to be the same.

We've explained the basics. What is coming next is more interesting – how to connect electrical loads.

Connecting electrical loads

You could have a couple of loads plugged in a circuit. In such a case, there are two main types of connection modes – in series and in parallel.

Connecting loads in series:

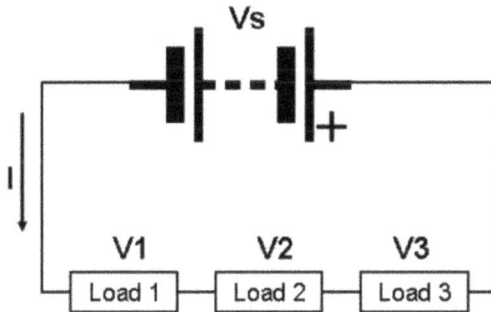

When loads are connected in series, the current flowing through each load is the same. The total voltage is a sum of the voltage drops in each load:

$$V_{total} = V_1 + V_2 + V_3$$

Connecting loads in parallel:

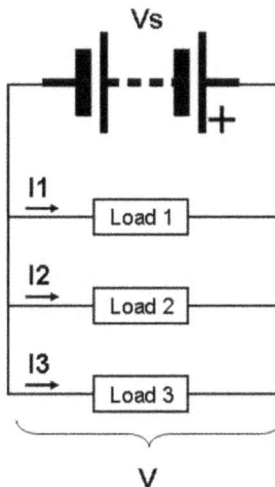

When loads are connected in parallel, the same voltage is applied to each of the loads. The total current in the circuit is a sum of the currents flowing through each load:

$$I_{total} = I_1 + I_2 + I_3$$

It should be noted that in both kinds of load connections, the total power remains the same:

$$P_{total} = V * I = P_1 + P_2 + P_3$$

Which connection is preferable – in series or parallel?

Connecting loads in series is not recommended for the following reasons:

1. If a load fails, current cannot flow through, and other loads cannot operate either.
2. Adding a new load is problematic. Since adding a new load means creating another voltage drop, the battery voltage might not be sufficient to power all the loads.

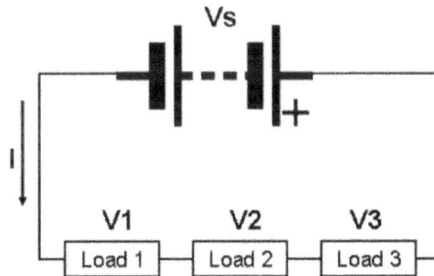

So, you need to replace the battery with a new one of higher voltage – otherwise neither of the loads will work!

Important!

Connecting loads in parallel is recommended, as:

- Each load is independent of the others. If a load fails, the other loads continue to operate.
- Adding more loads is not a problem – the other loads are not affected.

At your home and office, ALL the electrical loads are connected in parallel.

Loads as solar electric system components

Loads directly affect the performance of the solar electric system.

Extra loads can cause a system to fail if such loads require more power than the solar panels can generate or than the battery bank can store. Furthermore, the efficiency of the loads impact solar system's performance.

Therefore, the loads connected to a PV system should be as efficient as possible. Oversized or improper loads often result in system failure. Last but not least, loads powered by a solar electric system should not be operated unless they are needed.

DC loads

Lighting (incandescent or quartz halogen) and other resistive loads can be powered by solar generated electricity. The lamp and its fixture should be as efficient as possible. Using DC lighting equipment is a smart way to bypass any inverter inefficiencies related to DC to AC conversion.

"Heating" loads are not recommended to be powered by PV generated electricity. "Heating" loads comprise resistive heating appliances and tools such as toasters, coffee makers, soldering irons, room heaters and water heaters. Due to the high amount of energy they consume, such loads should be used only when there is no other option, or if the load will not often be used.

Inductive loads are appliances involving a motor or

an electromagnet. Many solar electric systems are designed to provide power to DC motors driving tools, fans, pumps, and appliances. The efficiency of the inductive load should also be as high as possible. DC motors are noted for their higher efficiency compared to AC motors.

Electronic loads, such as various audio-visual and communication electronic equipment, devices for data collection or security are typically DC-powered. These can be operated by solar electric systems with no problems. Many of these loads are sensitive to small variations in voltage. They usually are powered by battery-based photovoltaic systems, rather than grid-tied PV systems.

AC Loads

AC loads can only be used if the solar electric system includes an inverter. A good approach is to limit the number of AC loads as much as possible due to the energy lost in the DC-AC conversion performed by the inverter.

With photovoltaic systems, AC incandescent lighting should be minimized because of its poor efficiency, and for this reason, AC fluorescent lighting is recommended as more efficient. Furthermore, usage of AC appliances, such as toasters, dryers, soldering irons, and heaters should be minimized.

Inductive loads

Many appliances and power tools comprise motors operating on AC. Such devices typically require a "clean" source of AC power, i.e. an inverter of a pure sinusoidal waveform, which is more expensive. The reason for such a requirement is that motors operating by a not "clean" enough power source waste electrical energy. Such a wasted energy is dissipated through the motor housing as heat, which often reduces the motor's lifetime.

Another inductive type of load needing a "clean" AC power supply are microwave ovens. A "clean" AC power supply generates a sine-wave voltage.

Electronic loads

For certain electronic devices, such as communication equipment and small computers, providing AC power by a simple inverter is usually enough. Other electronic devices, such as video and audio equipment, require the use of advanced inverters.

Batteries are the solar system component that can be severely affected by the loads. If a solar electric system does not comprise a charge controller, oversized loads or excessive use of loads can quickly damage the battery bank, and it will have to be replaced soon due to overdischarging. Another adverse effect is overcharging which happens during times of low or no load usage at all, or extended periods of full sun. For these reasons, battery bank systems must be sized to match the loads connected to the system.

Automotive batteries are not designed for long time periods of low discharge, and this is the reason why they are not appropriate for, and should not be used in, photovoltaic systems.

Source:

1. Pop MSE, Lacho, Dimi Avram MSE. (2015-10-26).The Truth About Solar Panels: The Book That Solar Manufacturers, Vendors, Installers And DIY Scammers Don't Want You To Read, Kindle Edition. Digital Publishing Ltd.

2. Pop MSE, Lacho, Dimi Avram MSE. 2015. The Ultimate Solar Power Design Guide: Less Theory More Practice, Kindle Edition. Digital Publishing Ltd.

Solar electric systems and their components

There are two main types of photovoltaic (solar electric) systems – connected to the utility grid and disconnected from the utility grid.

Connected to the grid (grid-tied)	Not connected to the grid (off-grid)	
(with or without power backup)	Stand-alone: purely solar	Hybrid: solar with backup generator

Grid-tied systems allow you to offset a part of or all of your electricity demand to photovoltaics, thus reducing your electricity bills.

Most of the existing photovoltaic systems are connected to the local utility grid or are 'grid-tied'.

A typical grid-tied (also known as 'grid-connected', 'grid-direct' or 'grid-on') PV system does not provide electricity storage. Such a PV system generates electricity to provide a part of the energy needs of a building in daytime.

Grid-tied systems can be with or without an option for a power backup.

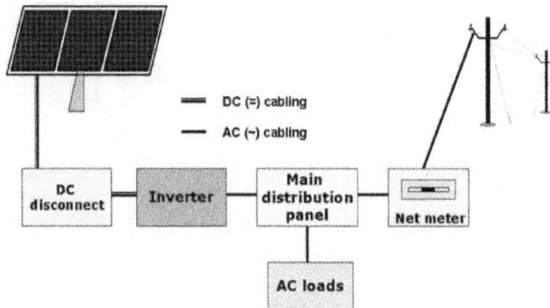

A simplified view of a grid-tied solar system without power backup

22

Here are the main components of a grid-tied system that does not provide any power backup:

- Photovoltaic array – generates DC electricity from sunlight
- DC disconnect – disconnects the solar array from the rest of the system
- Inverter – converts DC electricity into AC electricity
- Main distribution panel – the connection point between home electrical network and utility grid
- AC loads – the devices operating on AC electricity
- Net meter – measures the electricity imported from and exported to the utility grid.

The above picture does not show any Balance of System (BoS) equipment – all mounting and wiring systems and components needed to integrate the PV system in the existing infrastructure. The BoS equipment comprises various cables, jumpers, boxes, protection devices, etc.

When power generated by the PV system exceeds the building's energy needs, the surplus power is exported to the grid. This is called 'net-metering'. It provides you with the opportunity to get paid for the electricity you supply to the grid.

A grid-tied system does not operate during a power outage unless it has a power backup.

Grid-tied PV systems without power backup have the following advantages:
- Require almost no maintenance.
- Seasonal changes in solar radiation are not so important for their operation.
- Are easier to design and are less expensive than stand-alone systems.

In cases where frequent grid outages happen for relatively long periods, a **grid-tied system with power (battery) backup** is recommended:

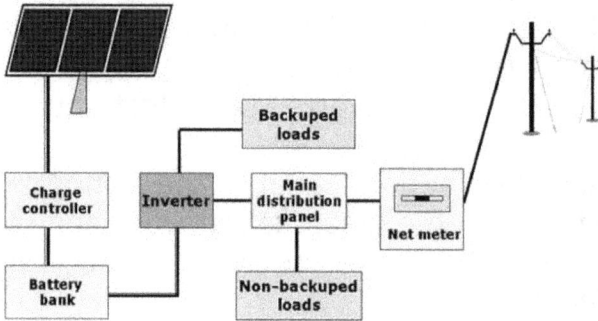

A simplified view of a grid-tied solar system with power backup

The components of such a system are:

- Photovoltaic array – generates DC electricity from sunlight
- Charge controller – regulates battery charging, thus increasing battery lifespan
- Battery bank – stores the electricity generated by the PV array
- Inverter – converts DC electricity into AC electricity
- Main distribution panel – the connection point between home electrical network and utility grid
- Backuped loads – all the AC and DC devices provided with power backup
- Non-backuped loads – those electrical devices that are not provided with power backup
- Net meter – measures the electricity imported from and exported to the utility grid.

Another type of solar systems is **off-grid** ones. They can be stand-alone (purely photovoltaic) and hybrid ones.

Off-grid solar electric systems are not connected to the grid.

Such systems are preferred in remote areas where buildings are far from any utility infrastructure. In such a case it is cheaper and easier to install a PV system to meet your daily electricity needs rather than pay for utility interconnection.

First, let's describe in brief stand-alone solar systems.

The simplest kind of a stand-alone PV system can be obtained by directly connecting a DC load to the PV array:

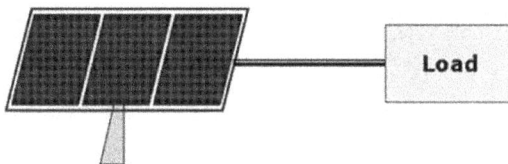

The load might be, for example, a DC fan or a DC water pump. Such devices use electricity right away after it is generated (i.e. in daytime), without any need to store it for later use.

Off-grid systems are usually provided with battery backup to store the generated energy if it is not used right away:

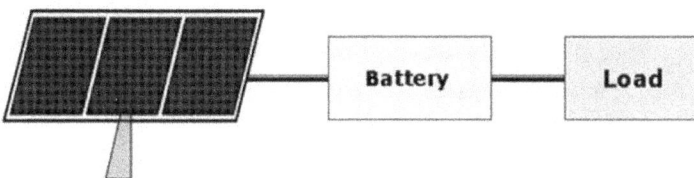

The load might be, for example, a TV set or a laptop. Since such devices operate not only in the daytime, a battery is needed to ensure their operation during night hours.

Here is an example of a stand-alone system designed to replace utility grid for remote buildings:

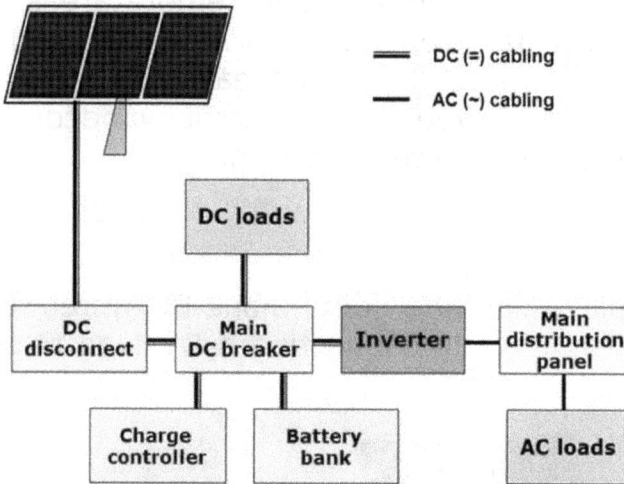

A simplified view of a stand-alone solar electric system

Here is a list of components of a stand-alone system:

- Photovoltaic array - generates DC electricity from sunlight
- DC disconnect – disconnects the solar array from the rest of the system
- Main DC breaker – connects the inverter to the battery and charge controller
- DC loads – all devices operating on DC power
- Charge controller – regulates battery charging, thus increasing battery lifespan
- Battery bank – stores the electricity generated by the PV array
- Inverter – converts DC into AC electricity
- Main distribution panel – the connection point between home electrical network and utility grid
- AC loads – all devices consuming AC power.

Stand-alone systems are 'photovoltaic-only' systems.

They contain no additional power generator apart from the solar array.

The second subtype of off-grid systems is **hybrid systems**.

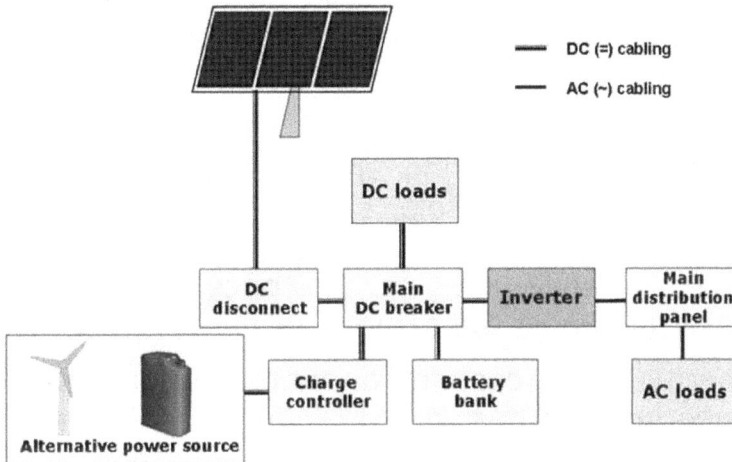

A simplified view of a hybrid off-grid system

A hybrid system is a stand-alone solar electric system with an alternative power source added – wind generator or fuel generator.

Hybrid systems are preferred in cases where too high energy consumption and/or long periods of cloudy days require a too bulky battery bank, which is expensive both to buy and to maintain.

Important!

Whether to choose a stand-alone or a hybrid system depends on:

- How much total daily energy you need.
- What kind of electrical applications you use – whether they are critical or not.

A hybrid system is suitable:

- If daily consumption of electricity is more than 2.5 kWh, or
- For regions with poor sunlight for long periods.

In every solar system, the solar panel energy generating capabilities are mainly responsible for solar electricity production and correspondingly for solar electricity production cost.

By carefully engineering and designing the rest of the components of a solar system, however, you can squeeze up to the 30% more power under same conditions, while keeping the price of your solar installation almost the same.

Furthermore, in case of an unprofessional system engineering and design, you may not only lose this additional 30%, but you may lose even more energy at the output of your system. Thus it would turn out that you have bought the most expensive solar panels on the earth! That is the primary mission of this book – to show you how to get or build the most efficient solar power system tailored for your needs.

Source:

Mayfield, Ryan. 2010. Photovoltaic Design and Installation for Dummies, Wiley Publishing Inc.

Solar photovoltaic modules

General information

Photovoltaic (solar) modules consist of photovoltaic cells connected in series and in parallel.

In a solar module, the solar cells are usually connected in series to provide a higher voltage.

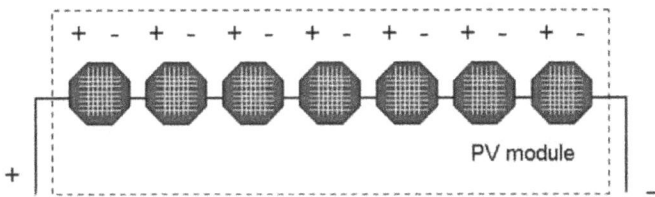

PV modules should be light and small enough to be installed on roofs, sometimes upon adverse conditions.

If you need to to get even higher voltage, you should connect PV modules in series. A group of solar modules connected in series is called '***string***':

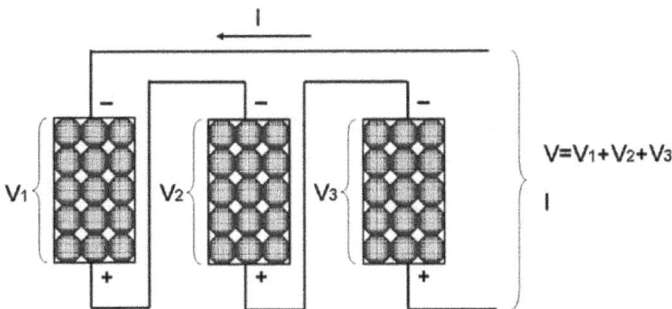

$$V=V_1+V_2+V_3$$

Efficiency is an important parameter of PV modules. Efficiency shows what part of the solar energy fallen onto a solar module's surface is converted into electrical power.

Important!

Every PV module has its rated power or peak power, denoted in kWp.

The peak power of the module, however, is not the real power the module can generate.

The actual power output of the module is always less than the rated power, due to the following factors:

- Manufacturer power tolerance
- Dirt and dust
- Temperature
- Cable losses
- Inverter efficiency
- Shading.

Manufacturer power tolerance is the percentage within which the manufacturers guarantee that the real power output will be the same as the rated power output. Such a percentage is never 100; the typical value is 95.

Dirt and dust cause losses when accumulated on the surface of a photovoltaic module. Dirt and dust particles could block the sunlight and thus reduce the power output. The content of dirt and dust in the air may vary with location and is usually the highest in an urban environment. In regions with heavy rainfall, dirt losses tend to be zero.

Temperature is one of the most important factors to be considered when designing a PV system. Temperature influences all the three most important electrical parameters of a PV module – voltage, current, and power. When the weather gets warmer, the output voltage and generated solar power goes down, and vice versa – when the weather gets colder, voltage and the generated power goes up. The solar cell temperature which determines whether the generated solar power goes up or down is 25°C (77F) regarding the Standard Test Conditions (STC).

Solar panels have a negative temperature coefficient which means that solar panel's performance declines while cell temperature increases.

The solar panel rated output power is defined under Standard Test Conditions (STC), which means:

- 1,000 W/m^2 of sunlight
- 25°C of cell temperature (77F)
- Spectrum at air mass of 1.5

Please, mind that generally for positive ambient temperatures the temperature of a solar cell is about 15°C (59F) higher than the ambient temperature, due to solar panel encapsulation.

Cable losses are inevitable in any PV system, especially when cables are long, which should be avoided whenever possible. A reasonably acceptable value of cable losses lies between 3% and 5%.

Inverter efficiency denotes what part of the input DC power is converted into AC power. The percentage is never 100, but values of inverter efficiency between 90% and 95% are widely assumed in practice.

Shading must be avoided, since even small shadows could severely reduce the performance of a PV module. A PV module consists of cells, and when shaded, any cell turns into a heat-dissipating resistor dramatically boosting the temperature of the PV module. This results not only in a sudden reduction in the output voltage but also in shortening the life cycle of the cells and modules. When mounted on

31

the roof, PV modules could easily underperform due to shading caused by trees, chimneys and other roof protrusions that are hard to eliminate.

PV modules differ mostly in their:

- Type – monocrystalline, polycrystalline, thin-film
- Power output (also known as 'power rating' or 'rated power') – between 10 Wp and 300 Wp
- Output voltage – 12 V, 24 V, 48 V or 60 V
- Size and weight – commonly 1.6 x 0.8 meters, or 5.25 x 2.62 feet.

Solar module types

Monocrystalline modules

Monocrystalline modules are the most efficient but also the most expensive ones. They come in blue or black color.

The less solar modules you need to produce a certain amount of power, the higher their efficiency. Generally, if there is not enough free space on your roof, you should choose modules of higher efficiency.

Polycrystalline modules

Polycrystalline modules are slightly less efficient and cost 30-50% less than monocrystalline ones when intended to produce the same amount of power.

Polycrystalline modules have a lifecycle of about 25 years. The practice has shown, however, that polycrystalline modules installed more than 25 years ago are still perfectly operational.

Polycrystalline modules are typically blue and can be easily distinguished by their multifaceted, kind-of-shimmering appearance.

Thin-film (amorphous) modules

Thin-film modules are the least expensive modules with the lowest efficiency – usually twice less than the efficiency of monocrystalline modules. Therefore, if you need to generate the same amount of power, you need twice more thin-film modules than monocrystalline ones.

Thin-film modules have a dark surface – usually colored in brown, gray or black. Thin-film modules are used in solar calculators.

Crystalline (mono- or poly-) PV modules are the most common type for home and business photovoltaic systems.

Crystalline modules come in a variety of size and shapes. The rectangular shape is the most common.

You should ask PV vendors to show you some solar installations to see how every PV array is matched to the roof and the house from an aesthetical point of view.

Photovoltaic modules have a lifetime of about 20-25 years, with about 80% of their rated power guaranteed within such a period.

Every solar panel has nominal power rated in 'watts-peak' (Wp) or 'kilowatts-peak' (kW). Here is a comparison between solar panel efficiency according to the area needed to install a solar panel of Nominal Power = 1 kWp:

PV cell material	Module efficiency	Area needed for 1 kWp
Monocrystalline silicon	13-16%	7 m² (75 sq.feet)
Polycrystalline silicon	12-14%	8 m² (86 sq.feet)
Amorphous silicon	6-7%	15 m² (161 sq.feet)

Thanks to the advancement in thin-film CdTe PV technology, however, some of the commercially offered CdTe thin-film photovoltaics have recently reached an average efficiency of 11.4%.

Copper Indium Gallium Selenide (CIGS) panels have a conversion efficiency of 10-18%. However, crystalline (mono- or poly-) PV panels are the most commonly used ones for home and business photovoltaic systems. Nowadays these solar panel types account for 90% of the photovoltaics market share.

In contrast, thin-film panels have about 10% market share, with amorphous silicon panels the most widely used among them.

Also, you should have in mind that during the period between the first six months and one year of operation, thin-film modules produce about 10-15% higher output than originally intended. After that, within a period of another six months to one year, the power yield settles down to the rated value sustained over the remaining years of operation.

Comparison between various solar module types

Monocrystalline modules

Price: High cost
Efficiency: Highest (13-16%)
Roof & mounting space required: Minimum space required; **7 m^2 (75 sq.feet) per 1 kWp installed power**
Ease of installation: Normal
Shading tolerance: Least tolerant
Type of plate glass: Predominately manufactured with tempered glass
Staebler-Wronski effect*: Suffer less
Technology age: Old & proven technology
Modules installed more than 40 years ago are still operational

Polycrystalline modules

Price: Lower cost
Efficiency: High (12-14%)
Roof & mounting space required: Less; **8 m^2 (86 sq.feet) per 1 kWp installed power**
Ease of installation: Normal
Shading tolerance: Slightly more tolerant than monocrystalline modules
Type of plate glass: Predominately manufactured with tempered glass
Staebler-Wronski effect*: Suffer less
Technology age: Relatively old & proven technology

Amorphous (thin-film) modules

Price: The lowest cost

Efficiency: Lowest (6-7% on average)

Roof & mounting space required: The most space-consuming: almost more than doubled space required than mono- and polycrystalline modules; 15 m^2 (161 sq.feet) per 1 kWp installed power

Ease of installation: Some modules need more mounting rails and more installation time, compared to mono- and polycrystalline modules. Thus installation cost and overall cost of your system increase

Shading tolerance: Highly tolerant

Type of plate glass: Predominately manufactured with plate glass (inferior to tempered glass)

Staebler-Wronski effect*: Suffer more

Technology age: Relatively new technology

*Staebler-Wronski effect results in a reduction of the solar module efficiency over time upon exposure to sunlight.

Source:

Pop MSE, Lacho, Dimi Avram MSE. (2015-10-26).The Truth About Solar Panels: The Book That Solar Manufacturers, Vendors, Installers And DIY Scammers Don't Want You To Read, Kindle Edition. Digital Publishing Ltd.

Connecting **photovoltaic modules**

To produce more power and energy, photovoltaic modules are connected in series or in parallel. Thus, they form a photovoltaic array. To avoid loss of power in a PV system, you should only connect solar modules of the same type.

Connecting PV modules in series means joining the positive terminal (+) of a module to the negative terminal (-) of the next module, joining the negative terminal of that module to the positive terminal of the next module, and so on:

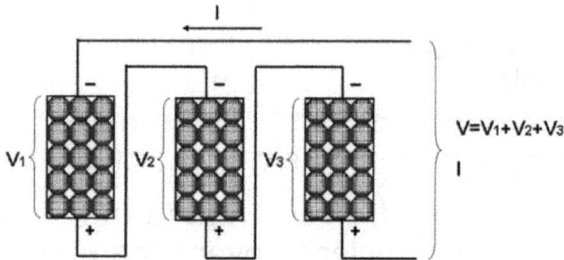

$$V = V_1 + V_2 + V_3$$

A set of PV modules, connected in series, is known as 'string'.

Solar modules are connected in series to obtain a higher output voltage. The maximum system voltage, however, must not be exceeded.

For modules, connected in series, the total power is calculated as follows:

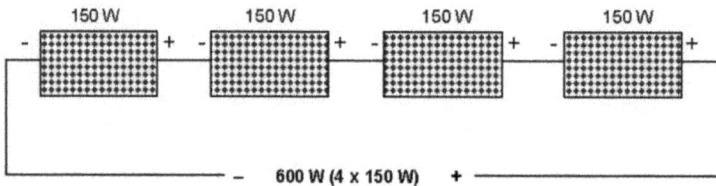

150 W   150 W   150 W   150 W

600 W (4 x 150 W)

Total connected power = 150W + 150W + 150W + 150W = **600W**

38

Important!

If among modules connected in series, a module has rated power lower than the rated power of the other modules, then that module will drag down the output of the whole system:

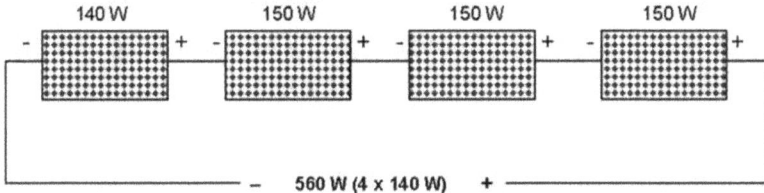

| 140 W | 150 W | 150 W | 150 W |

560 W (4 x 140 W)

The reason for this is the same current flowing through every load in a series connection. Therefore, the current of a string is defined by the current of the "weakest" module in it. If the current of the module of lower rated power is lower than the current of the other higher power modules, it will drag down the performance, i.e., the power of the remaining modules.

We advise you to carefully consider these issues when planning to buy and install second-hand solar modules!

Advantages of connecting PV modules in series:

- PV modules are easier and quicker to mount
- Shorter cables are needed
- Smaller cable cross-sections are needed
- Power losses are reduced

Important!

Connection in series is preferred when a solar array is *NOT IN SHADE*.

Series connection is the most used pattern for connecting solar modules. The main drawback is the occurrence of high DC voltages (200-600 Volts; remember summing up voltages for series connection of loads!), which are extremely dangerous for humans.

The series connection of solar modules is preferred in grid-tied PV systems.

PV modules are connected in parallel in the following way:

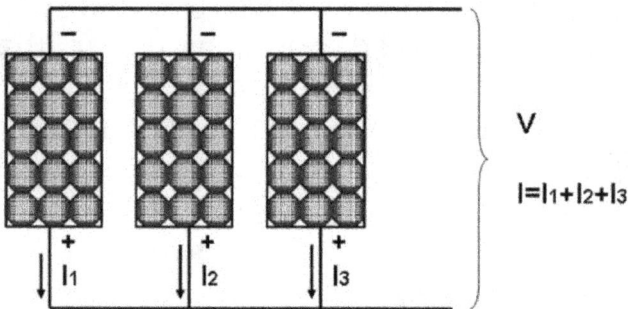

$$V$$
$$I = I_1 + I_2 + I_3$$

That is, positive (+) terminals of the modules are connected together, and negative (-) terminals of the modules are connected together.

Solar modules are connected in parallel, in order to obtain higher output current. For PV modules connected in parallel, the total power is calculated as follows:

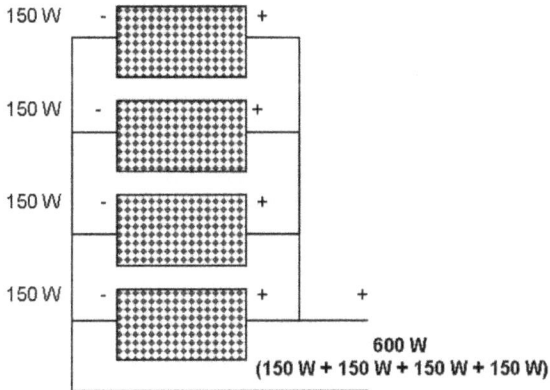

Total connected power = 150W + 150W + 150W + 150W = **600W**

Unlike connection in series, if among modules connected in parallel there is a module of power output lower than the power output of the other modules, this **does not seriously affect** the total power output of the array, provided that such a module is of the same rated voltage as the other panels.

Otherwise, it will drag down the voltage of the rest panels, thus reducing their output power, which in turn will reduce the output power of the whole string:

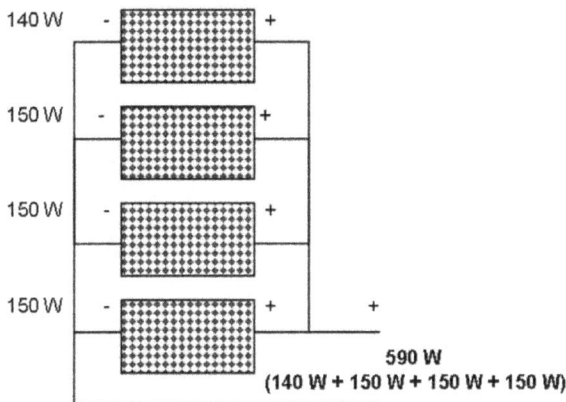

41

Maximum voltage on a string of modules must always be lower than the maximum input DC voltage of the inverter (read the inverter section for more details).

Important!

Parallel connection of solar modules is preferred when a PV array is *IN SHADE*.

Parallel connection is preferred in stand-alone PV systems.

Examples of connected solar modules

As mentioned in the section 'Electricity Basics,' solar modules are commonly generators rather than loads. Nevertheless, the same above principles are valid for series and parallel connections:

- For series connection, the current flowing through each load is the same, while the total voltage on all the loads is the sum of the individual voltages on each load.
- For parallel connection, the voltage on each load is the same while total current is the sum of the individual currents flowing through each load.
- Regardless whether connected in series or parallel, PV modules produce the same amount of power.

Mixed connection is also possible, with the same principles mentioned above being valid.

Example 1) Series connection of two modules

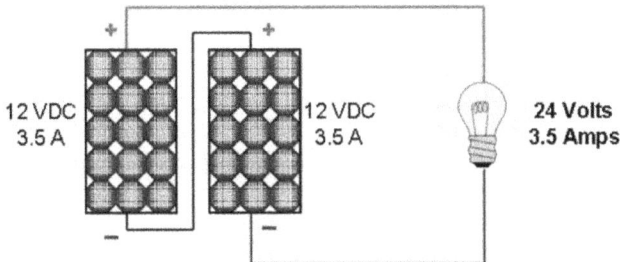

Total voltage = 12V + 12V = 24 Volts
Total current = 3.5 Amps
Total power output = 24 Volts * 3.5 Amps = 84 Watts

Example 2) Parallel connection of two modules

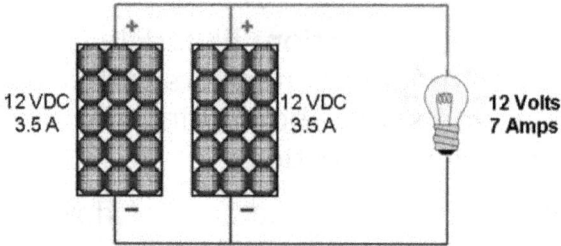

Total voltage = 12 Volts
Total current = 3.5A + 3.5A = 7 Amps
Total power output = 12 Volts * 7 Amps = 84 Watts

Example 3) Series connection of six modules

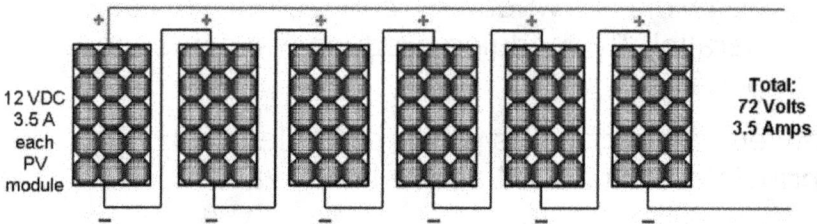

Total voltage = 12V + 12V + 12V + 12V + 12V + 12V = 72 Volts
Total current = 3.5 Amps
Total power output = 72 Volts * 3.5 Amps = 252 Watts

Example 4) Parallel connection of four modules

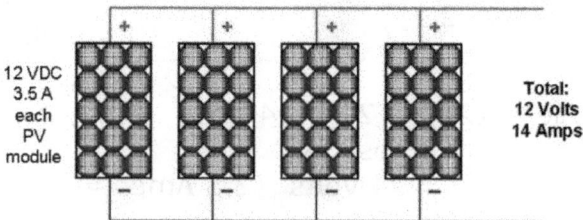

Total voltage = 12 Volts
Total current = 3.5A + 3.5A + 3.5A + 3.5A = 14 Amps
Total power output = 12 Volts * 14 Amps = 168 Watts

Example 5) Mixed connection of four modules

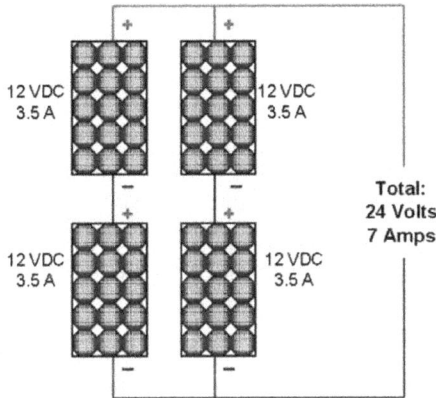

Total voltage = 12V + 12V = 24 Volts
Total current = 3.5A + 3.5A = 7 Amps
Total power output = 24 Volts * 7 Amps = 168 Watts

Example 6) Mixed connection of eight modules

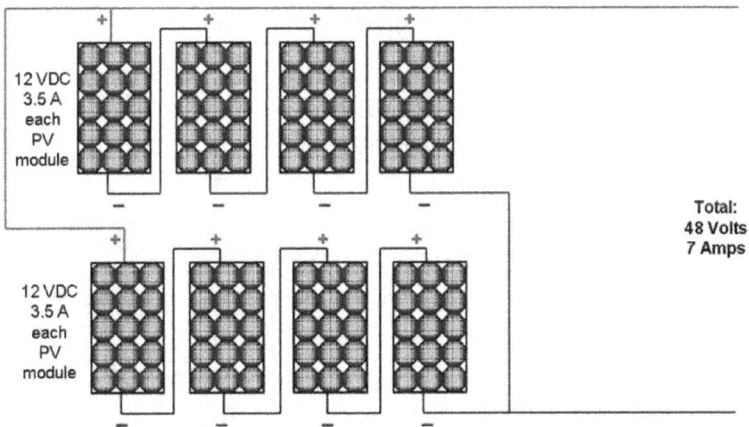

Total voltage = 12V + 12V + 12V + 12V = 48 Volts
Total current = 3.5A + 3.5A = 7 Amps
Total power output = 48 Volts * 7 Amps = 336 Watts

Connecting solar modules in summary

Important!

- Connecting solar modules together (whether in series or in parallel) allows your solar panel system to produce more power, and hence generate more solar electricity.
- Connecting solar modules in series allows the solar array to run at a higher voltage. Such an approach is often used in grid-tied systems to produce more power.
- Connecting solar modules in parallel allows the solar array to produce more power and maintain a lower voltage on a single module. This approach is often used in stand-alone systems operating on lower voltage, due to the limited voltage range of the battery bank.
- When connecting PV modules **in series**, always use identical modules from the same manufacturer, as the overall performance of the array will be limited to the performance of the lowest performing solar module in the string.
- As a last resort, if you, for whatever reason, have to use solar panels from different series by the same manufacturers or by a different manufacturer, please consider using a separate charge controller or inverter, depending on the solar panel system type you have selected. The higher yield of the solar panels will quickly justify the money invested in such extra hardware.
- Carefully consider carefully the type of PV modules in advance and do not mix different modules from different manufacturers.

Source:

1. Pop MSE, Lacho, Dimi Avram MSE. 2015. The Ultimate Solar Power Design Guide: Less Theory More Practice, Kindle Edition. Digital Publishing Ltd.
2. Pop MSE, Lacho, Dimi Avram MSE. (2015-10-26).The Truth About Solar Panels: The Book That Solar Manufacturers, Vendors, Installers And DIY Scammers Don't Want You To Read, Kindle Edition. Digital Publishing Ltd.

General rules for wiring solar panels

When connecting solar panels, please be sure to comply with the following rules for solar panel wiring:

- Use cables that are sized correctly (refer to **"The Ultimate Solar Power Design Guide: Less Theory More Practice" for sizing guidelines and formulas**)
- Earth the solar array correctly: connect the frame of each solar panel to an earthing cable and connect this cable to the central earthing terminal
- Junction boxes should be well sealed to prevent corrosion.
- In case of a solar array consisting of more than one solar panel, carefully consider the maximum system voltage, and calculate the number of panels that can be wired in series and parallel (refer to **"The Ultimate Solar Power Design Guide: Less Theory More Practice"**).

Bypass diodes and fuses

A group of PV cells connected in series forms a cell string. Each cell string is protected by a bypass diodes or fuses.

A diode is a piece of semiconductor allowing the current to flow in one direction only.

Bypass diodes are placed in a PV module or a PV string to prevent reverse voltage.

Reverse voltage occurs when one or more modules in a string are partially shaded (by a leaf or any other obstacle). Lack of bypass diodes could cause:

• Undesired heating with subsequent damage of the PV module
• Reduction of the array current, with subsequent power loss.

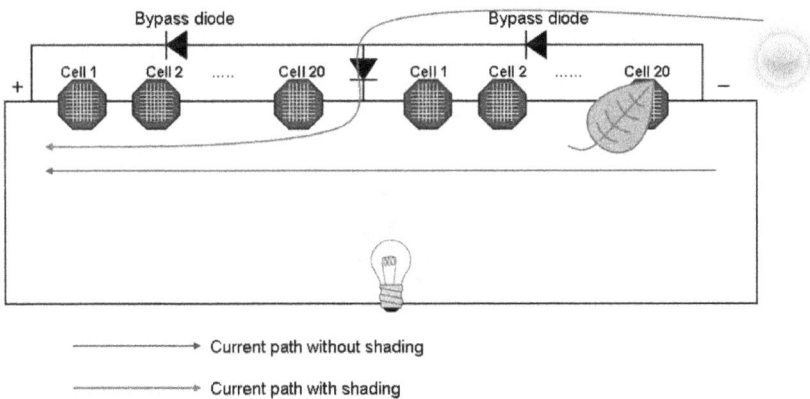

Bypass diodes are usually integrated into PV modules upon their manufacturing. These factory-installed, built-in bypass diodes provide a proper circuit protection for the systems within the specified system voltage so that no additionally installed individual bypass diodes are needed.

49

Bypass diodes are connected in parallel with cell strings, and typically for PV modules that are on the market, a bypass diode is shared between 20 and 24 solar cells. Bypass diodes are installed in the solar module's junction box.

Important!

If a solar array includes bypass diodes, they can be tested by shading the module under test and then checking, by an ammeter, whether any current is flowing through from the rest of the array. If there is little or no current flowing, the diode should be replaced.

Fuses are used for the same purpose as blocking diodes (see *Components of stand-alone (off-grid) systems, Blocking diodes*) – to protect the cables from overcurrent. Fuses can be designed in various shapes and sizes.

Fuses are used when a large number of PV strings are connected in parallel and the cable's rated current in one string should be below the Isc, i.e. the short circuit current of all the PV modules connected together. In case of short circuit in a string, as a result of module failure or shading, the remaining strings should continue to operate unaffected.

If no fuses are used, the price to be paid is using cables of rated current greater than the short circuit current of the PV array.

Integrated bypass diodes have one clear disadvantage.

When faulty built-in bypass diodes are to be replaced, this is only to be done by the manufacturer. Luckily faulty bypass diodes are rarely to occur. Such a situation might happen, if:

- Solar modules are frequently shaded
- Modules with the wrong polarity are connected
- The outdoor part of a PV system is struck by a lightning strike.

Usually, Schottky diodes are used as bypass diodes. The main advantages of a Schottky diode are:

- Lower forward resistance leading to a very low forward voltage drop, compared to other types of diodes
- Faster switching capabilities than other types of diodes.

In practice, all the three above causes for bypass diodes failure can be minimized.

Solar array maintenance

- For maximum performance, clean your solar panels at least one per year.
- Check whether the current of the solar array and the loads connected does not exceed the pertaining rated values.
- Make sure that all terminals are tightened; also check for any loose or broken wire connections. No loose wires must be in contact with array terminals.
- Clean the solar panels by using your hand, paper towels, a soft cloth or a sponge. If necessary, use a mild soap or a non-abrasive cleaner.

How often you should clean your panels depends on the environment.

For solar array near a road cleaning once or twice a month is recommended. Regularly cleaning the solar panels is important, since dust and dirt directly affect the system performance – a soiled panel allows less sunlight to reach to solar cells thus reducing the solar panel's total output by 15 to 25%.

A combination of dust/dirt and individual leaves fallen on the panel (to say nothing at shading by trees!), shading particular solar cells, is even more detrimental – it can reduce the solar output by 50 to 75%! Therefore, not only regular cleaning but also regular inspection is recommended.

Most solar panels are made with glass that is durable enough to withstand high winds or hailstones. Nevertheless, panels can be broken by stones or damaged if dropped. A broken solar panel is usually impossible to repair. You should mind that if only one solar cell is broken, the panel becomes useless.

Source:

Antony, Falk, Christian Durschner, Karl-Heinz Remmers. 2007. Photovoltaics for Professionals: Solar Electric Systems Marketing, Design and Installation, Routledge.

Components of grid-tied PV systems

Grid-direct (on-grid, grid-tied, grid-connected) photovoltaic systems:

- Produce electricity
- Use electricity from the grid
- Export electricity to the grid

Grid-tied systems can be designed with or without a battery backup.

Grid-tied systems without battery backup are built in regions where power outages happen rarely and have a short duration. Here are the main components of a grid-tied system without battery backup:

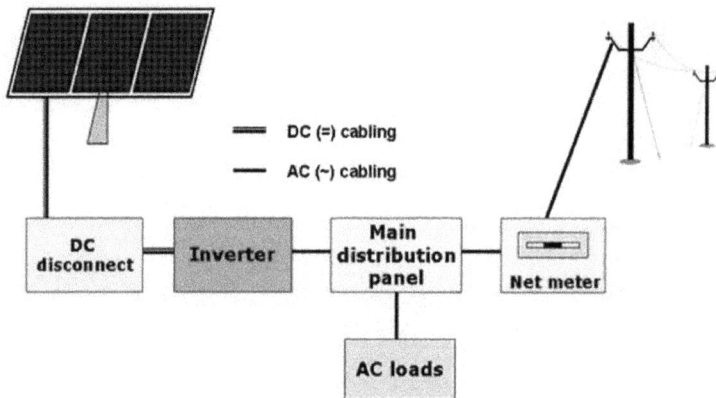

- Photovoltaic array - generates DC electricity from sunlight
- DC disconnect – disconnects the solar array from the rest of the system
- Inverter – converts DC electricity into AC electricity

- Main distribution panel – the connection point between home electrical network and utility grid
- AC loads – the devices operating on AC electricity
- Net meter – measures the electricity imported from and exported to the utility grid.

In case of an outage, a grid-tied PV system shuts down until the utility is up again. Shutting the PV system down prevents technicians that might be doing certain repair works on the utility grid at a certain moment from getting an electric shock.

Grid-tied systems with battery backup are preferred in areas where power outages happen more often and by users for whom electricity outage is not an option even for short periods. A grid-tied system with battery backup comprises the following components:

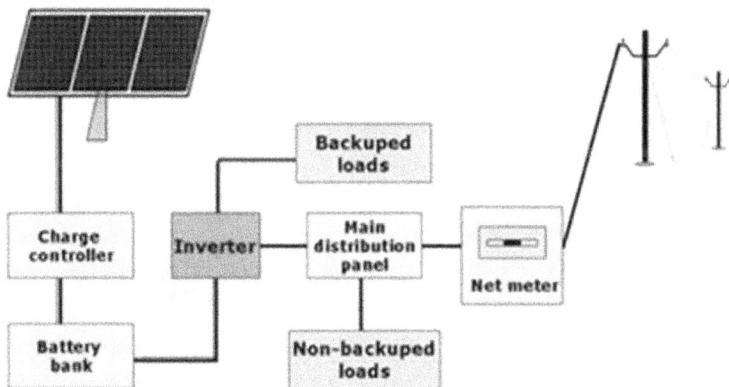

- Photovoltaic array – generates DC electricity from sunlight
- Charge controller – regulates battery charging, thus increasing battery lifespan
- Battery bank – stores the electricity generated by the PV array

- Inverter – converts DC electricity into AC electricity
- Main distribution panel – the connection point between home electrical network and utility grid
- Backuped loads – all the AC and DC devices provided with power backup
- Non-backuped loads – those electrical devices which are not provided with power backup
- Net meter – measures the electricity imported from and exported to the utility grid.

If you have a grid-tied system, you use the electricity generated from the system during the day while the sun is shining. After the sun goes down, your home network automatically switches to using electricity from the grid. Thus you need to pay the utility for the electricity provided during night periods only.

Certainly, you use electricity from the grid when electricity generated by your PV system does not fully cover your household electrical consumption.

If the photovoltaic generator produces more electricity than you consume, the 'excess' of electrical energy is exported to the grid and you get paid for that.

Grid-tied systems are less expensive and require less maintenance than stand-alone (off-grid) systems.

An obvious disadvantage is that grid-tied systems shut down in case of power outage. This drawback can be avoided by buying a grid-tied system with a battery backup option.

Inverters for grid-tied systems without battery backup

The primary functions of a grid-tied inverter are:

- Converting DC electricity produced by solar array to AC electricity with sine wave voltage and frequency in compliance with your local standard.
- Disconnecting your solar electric system from the grid during grid power outage to ensure safety repair works being performed on the grid.

Important!

A grid-tied inverter can be integrated into a PV array.
The inverter is not always an individual block located outside the PV array.
The inverter might be physically integrated into the PV modules so that they can be directly connected to the utility grid.

Since every grid-tied inverter stops working during the grid outage, you do not have any electricity during this outage as well. This feature is called `anti-islanding protection`.

If you want to have power supply after the grid fails, you should use grid-tied systems with power backup (refer to section 'What makes grid-tied systems with power backup attractive?'). Grid-tied systems with power backup use a different kind of inverters. For more details, refer to section '*Inverters for grid-tied systems with power backup*'.

A grid-tied inverter is connected directly to the PV array at one side and the utility at the other side.

Such an inverter does not provide any energy storage features.

Grid-tied inverters are highly efficient and straightforward to install.

Grid-tied inverters are also called utility-interactive, since they cannot operate when the grid is off. Their operation is always in parallel with the utility in providing power for home and office electrical equipment. Moreover, they are designed so that they can simultaneously send power back into the utility grid.

In small home PV systems the solar array is usually connected to a **single (central) inverter**:

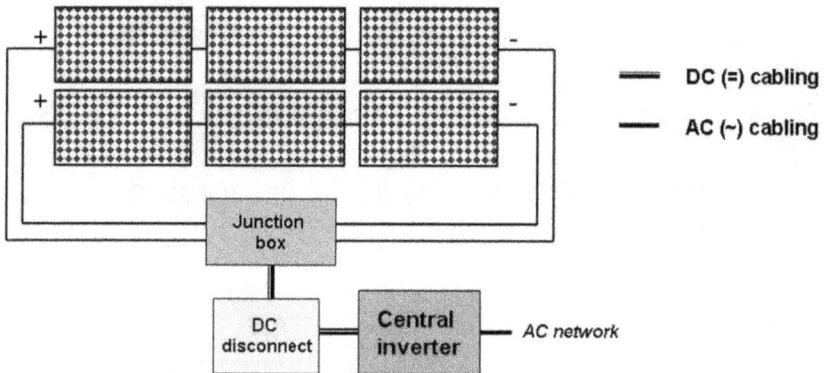

Source:

Antony, Falk, Christian Durschner, Karl-Heinz Remmers. 2007. Photovoltaics for Professionals: Solar Electric Systems Marketing, Design and Installation, Routledge.

Multiple inverters are used:

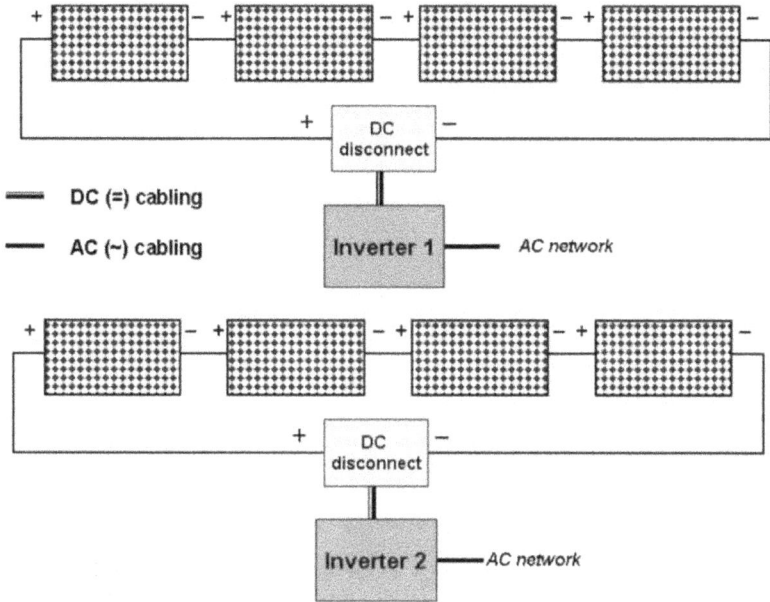

- For larger PV arrays – to prevent total system breakdown when the single inverter fails.
- If a part of the PV array is shaded, and this cannot be avoided.
- If parts of the PV array have a different orientation or different tilt angles.

Source:

Antony, Falk, Christian Durschner, Karl-Heinz Remmers. 2007. Photovoltaics for Professionals: Solar Electric Systems Marketing, Design and Installation, Routledge.

Microinverters

Microinverters (also known as *module inverters*) have an output power of less than 250 W. A microinverter is a part of a PV module and operates as a central inverter, with the exception that it converts DC into AC power only for the solar module to which it is connected.

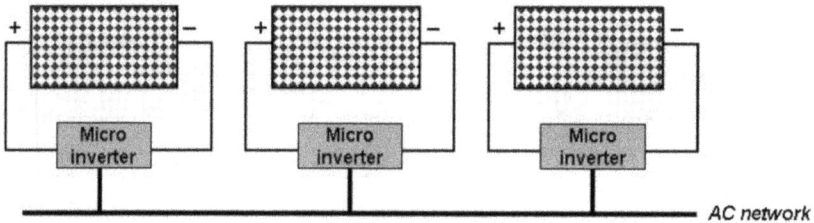

AC network

The following is typical about microinverters:

- Usually, the PV module and the inverter form a common unit – called 'module inverter' or 'AC module'.
- Each module has its inverter.
- The inverter can be either mounted by the manufacturer in the PV module's junction box or attached to the PV module.
- They are suitable for façade-integrated systems or systems with increased risk of PV array shading.

Benefits of microinverters:

- Easy and simple to install.
- Allow mounting PV modules on different surfaces and facing different directions.
- If one module gets shaded, the operation of the remaining modules is not affected.
- Safer than central inverter – there is no high voltage DC cabling between modules.

- Every inverter can achieve PV module's optimal performance at its maximum power point (MPP).
- Allow easy extension – the PV system can be only enhanced by a single module if needed.

Drawbacks of microinverters:

- Relatively expensive
- Have shorter lifecycle than PV modules
- Less resistible to heat
- In case of inverter failure, the whole module has to be replaced.

Microinverters can also be successfully used in large-scale systems by providing the following advantages:

- There is no need for DC cabling.
- Shading of a module and/or inverter failure cannot affect the rest part of the PV array.
- A perfect choice for PV systems containing PV modules of different power tolerances.
- The voltages used are lower than 120 V and thus are less dangerous.

Source:

Antony, Falk, Christian Durschner, Karl-Heinz Remmers. 2007. Photovoltaics for Professionals: Solar Electric Systems Marketing, Design and Installation, Routledge.

Maximum Power Point Tracking

At present most grid-tied inverters are provided with Maximum Power Point Tracking (MPPT) ensuring up to 20÷30 % more solar power generated, by delivering an optimal load to the solar array.

Constantly changing solar irradiation and ambient temperature causes constant changes in electrical characteristics of the solar modules. Hence the solar array needs different optimal loads to deliver an optimal power to the remaining units of the solar system.

Such inverters are known as MPP-tracking (MPPT) ones. The extra 20-30% of the solar power provided by the array, however, not only compensates the higher price in the long run but can bring you more savings as well.

Important!

If you use an MPPT inverter in your grid-tied system, you must be sure that the maximum and minimum output voltages of your solar array fall within the maximum power point tracking window of the inverter:

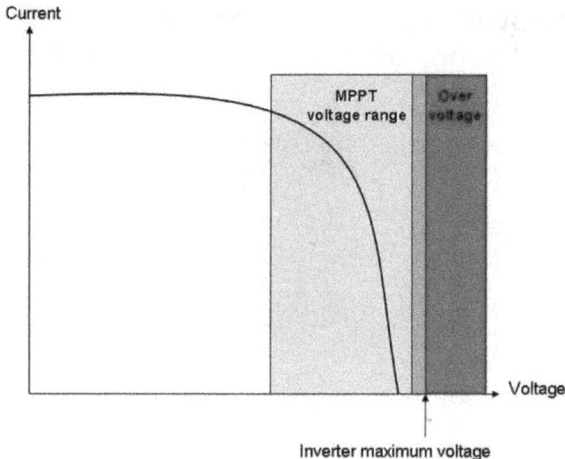

Current

MPPT voltage range

Over voltage

Voltage

Inverter maximum voltage

Such a tracking window is defined by maximum and minimum input tracking voltages of the inverter.

A solar array reaches its maximum output voltage at the lowest ambient temperature, while it reaches its minimum output voltage at the highest ambient temperature.

By not matching the inverter's tracking window, at extreme ambient temperatures an MPPT inverter will stop tracking, and the efficiency of your solar system is going to degrade.

Important!

It is important to ensure that:

- The minimum output voltage of the solar array does not fall below the minimum input voltage of the inverter. Otherwise, the inverter will not operate correctly.

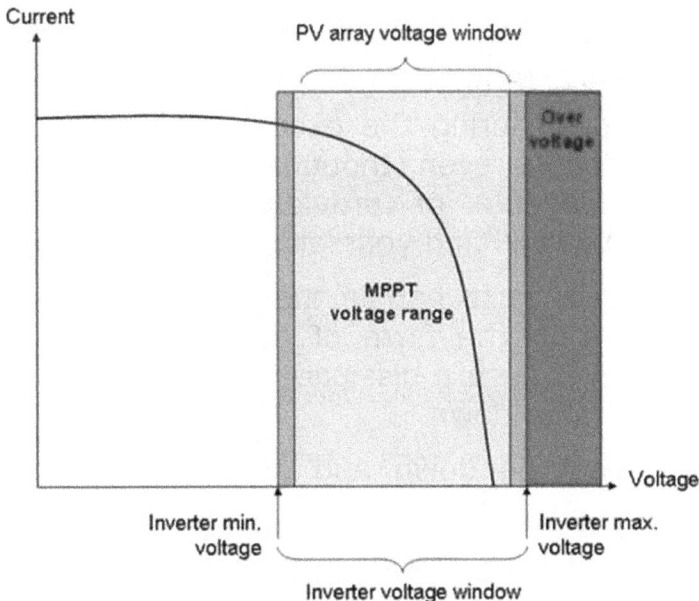

- The maximum output voltage of the solar array is always below the maximum input voltage of the inverter. The inverter can get damaged, should its maximum input voltage be exceeded.

So, if you assemble your solar system yourself, please do check this tracking window either with the solar installer or with the inverter supplier.

What is more, the inverter should be sized properly and matched to the AC loads as well.

If the inverter is overloaded or underloaded, its efficiency decreases. A lower efficiency means that in the long run you will lose solar power, and hence money saved on electricity bills.

On the other hand, the inverter power rating should be chosen so as to allow the scalability and growth of your solar power system.

Important!

If the inverter's power rating is significantly lower than the power required by your loads, the inverter will shut down during the excessive power request from the loads, even though your solar module might be capable of providing such a power. Therefore, you will find yourself in a power outage.

The inverter tries to handle the excessive power by dissipating it in the form of heat, which leads to overheating. If such a dissipation is not enough, the inverter will shut down.

Frequent shutting down and overheating decrease the lifetime of the inverter.

It is important to mount the inverter in a well-ventilated place to ensure optimal cooling. Otherwise, the inverter will try to deal with self-heating by reducing generated AC output power to protect itself until overheating reaches its shutting-down point.

Source:

Mayfield, Ryan. 2010. Photovoltaic Design and Installation for Dummies, Wiley Publishing Inc.

Ground-fault protection and safety

Ground-fault protection is provided to every PV system by the inverter. Ground-fault protection is used for fire prevention in case of fault in any part of the wiring.

A ground fault occurs when a current-carrying conductor comes into contact with the frame or chassis of an electrical device. All system components, including equipment boxes and PV mounting equipment, should be grounded.

Grounding means to take one conductor from a two-wire system and connect it to the ground. In a DC system, the conductor that should be connected to the ground at a single point is the negative one. The grounding of the PV system is always implemented inside the inverter, and not at the PV array.

Regardless of who is going to install the inverter, the following rules are to be kept to ensure maximum safety:

Important!

- The inverter should be mounted high enough above the ground to eliminate access to it by children and animals.
- Every inverter should be installed in a dry and well-ventilated place, with enough space for heat dissipation.
- The inverter, along with its cabling, must be kept away from any flammable materials – gas, oils, solvents and other volatile substances that can easily be ignited by an accidental spark produced by the inverter.

Mounting the inverter in a **small room**, with **poor ventilation**, near flammable materials, can result in **fire and even explosion**!

Inverter requirements & specifications

Important!

The essential requirements of grid-tied inverters:

- Generation of sinusoidal voltage synchronized with the sinusoidal voltage of the grid
- Precise tracking of the maximum power point of the I-V curve of the PV generator
- Reliable operation at both high and low temperatures
- Overcurrent protection
- Waterproof enclosure – grid-tied inverters are usually intended for outdoor installation
- Compliance with the relevant national standards and regulations.

Important!

The primary specifications of grid-direct inverters:

- Rated input DC power – usually selected 20% lower than PV array peak power, due to solar array losses. However, it should be checked whether the inverter rated input power is higher than the power the solar array is expected to produce at a lower ambient temperature.
- Rated input DC voltage – typically between 75 V (minimum value) and 750 V (maximum value) for most inverters used in residential grid-tied systems. The PV array output voltage should fall within this voltage window. Otherwise, the

inverter either will not work or will be damaged. Rated input DC voltage of the inverter should be about 20% higher than the solar array peak voltage.

- Maximum input DC current – should always be higher than the short-circuit current of the PV array.
- Output voltage – usually 120 VAC or 240 VAC for most residential buildings.
- Output frequency – 50Hz in Europe, 60 Hz in the USA.
- Efficiency – describes the percentage of losses arising after DC to AC conversion.

How to select the inverter for your grid-tied system?

Important!

Here are the essential technical criteria for selecting the inverter for your grid-tied system:

1) The output power of the inverter should be (0.9÷0.95) of the solar array peak power.
2) The maximum voltage of the PV array should be lower than the inverter's maximum input DC voltage (otherwise, the inverter will get damaged).
3) The minimum allowable voltage of the inverter should be less than the minimum DC voltage of the PV array (otherwise, the inverter will not work).
4) The working voltage range of the PV array should be within the inverter's MPP voltage range (otherwise the inverter will not track the PV array correctly and the PV system will underperform).
5) The maximum current of the PV array should be below the inverter's maximum input DC current (otherwise the inverter will get damaged).

Inverters for grid-tied system without battery backup in summary

You should remember the following about grid-tied inverters:

- They only operate when the utility is on.
- When the utility goes down, grid-tied inverter turns off immediately. This process is a part of standard compliance requirements and is called 'anti-islanding'. It prevents technicians doing any repair works from getting an electric shock, should the inverter keep sending power to the grid. Therefore, whether the sun is shining or not, your home or office will be in a power blackout.
- Most grid-tied inverters are based on Maximum Power Point Tracking (MPPT) – a feature 'squeezing' maximum possible amount of power from the PV array.
- The inverter is connected to the utility grid either directly or via the building's electrical system:
  - In case of a direct connection, the generated AC electricity is sent towards the utility grid.
  - In case of a connection via the building's grid, the AC power generated by the PV system is first consumed in the building, and what remains unused, is directed to the utility grid.
- Three different types of inverters are currently available on the market – sine-wave, quasi (modified) sine-wave and square-wave ones.

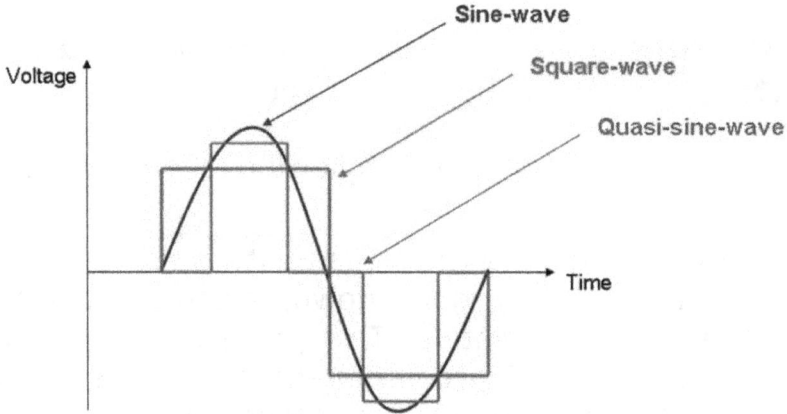

Certain electronic devices, such as mobile phones, microwave ovens, computers, vacuum cleaners, etc., might have problems while operating with a quasi sine-wave inverter.

Furthermore, quasi sine-wave inverters may create additional noise to audio and TV equipment. Inductive loads, such as fridges, pumps, drills, etc., must be powered by pure sine-wave inverters.

A square-wave inverter is of less quality than a modified sine-wave inverter.

Important!

Although the most expensive ones, **sine-wave inverters are the only possible choice for any grid-tied system** – not only because they are suitable for any applications, but also because they comply best with the applicable regulatory requirements.

Other components of grid-tied systems

Net metering: utility meter

Net metering is selling excess energy back to the local utility company usually at the same rate you are charged, while you are drawing energy from the utility.

For example, if your solar panels produce more electricity during the day than your household needs, the surplus gets sold back to the utility company at the retail rate. In the evening, when your household needs to draw on the utility company for electrical power, you pay the same retail rate for the electricity you use.

If the amount of electricity you sell is equal to the amount of electricity you have to buy, your "net usage" would be zero. Therefore, net metering balances energy you sell against energy you buy.

As a consequence, you either pay the difference or accumulate credits you can use to pay for electricity later on.

In a grid-tied system the utility meter tracks both electricity coming in from the grid, and excess electricity (produced by the solar panels) going back to the grid.

The dials will spin backward when such excess electricity is being sent to the grid, and the homeowner will receive credit for it if the utility participates in net metering.

In grid-direct systems, usually a bidirectional 'smart'

meter is used, enabling tracking of both imported and exported energy. In any case, however, requirements and regulations related to selling electricity to local utility company should be carefully investigated.

Balance of System Equipment

Balance of System (BoS) equipment includes all the mounting and wiring equipment necessary to integrate the PV system into the building's construction and electrical network. Such equipment includes array circuit wiring, fusing, inverter AC and DC disconnects, overcurrent protection and ground-fault protection.

As a rule fuses, connectors, surge arrestors, and circuit breakers are mounted in one or more junction boxes that are usually located on the periphery of the PV modules, and are designed to withstand environmental temperatures and ultraviolet radiation.

Fuses

String fuses are used to prevent, in case of string failure, too high currents flowing along other solar strings connected in parallel. The overcurrent protection is used to provide protection against internally induced voltages in the PV system. Such induced voltages caused by nearby lightning strikes can damage the solar equipment.

Surge protectors/arrestors

Surge protectors/arrestors are used to direct the currents arising upon lightning strikes towards the ground, since all the electronic components in a PV system (PV array, blocking and bypass diodes and inverter) need protection from high currents and voltages. Surge protectors are designed to withstand high voltages (up to 6 kV) and high currents (at least 5 kA).

Disconnects (circuit breakers)

Disconnects are used for disconnecting the indoor part (the inverter, the loads, and the utility meter) from the source of power which is the outdoor part (the PV array). The PV array cannot be switched off – it either produces electricity in the daytime when the sun is up or generates no electricity at night.

A disconnect can be either an individual device or integrated into the inverter itself. Sometimes it could be placed in the PV generator junction box.

The purpose of a DC disconnect is disconnecting the link between the PV array and the inverter. PV modules always generate some voltage when exposed to sunlight, and sometimes for short periods – during installing, repair or maintenance works – the PV array should be disconnected from the inverter.

As was mentioned above, DC disconnect could be integrated into the junction box. It should be noted, however, that while junction box might not be necessary, DC disconnect is a must.

For safety reasons, in case the DC switch is an individual device, it should be located indoors, usually immediately before the inverter.

AC disconnect is used to separate the household inverter and the loads connected to the PV generator from the utility grid.

Junction (combiner) boxes

If a PV array comprises a couple of strings, you need a junction (combiner) box to connect all the cables that will further connect to the inverter. Junction boxes are located on the back of the PV modules and are used for:

- Connecting cables coming in parallel from the strings of modules
- Providing room for mounting string fuses
- Providing room for mounting overcurrent protection
- Simplify testing of the module strings.

Combiner boxes are typically located on the roof. A combiner box should be provided with a double insulation and should allow laying positive and negative cables separately.

In case of multiple inverters, each inverter requires a separate junction box. The junction box could also integrate the main DC disconnect. In such a case, however, the box should be easily accessible.

Sometimes junction box might not be needed – for example, when PV group only comprises a couple of solar modules.

Cables

Cables (wiring, conductors) are used to connect the individual components of a PV system. Cable is a wire put inside a conduit or a pipe for protection.

A distinction should be made between DC and AC cabling in a PV system.

The DC part of the cabling comprises the outdoor

laid cables and the wiring between the modules (gathered in junction boxes), between the strings (gathered in combiner boxes) and the connection to the inverter.

To provide protection against ground-faults and short-circuits, positive and negative poles of the cables should always be separated from each other. Furthermore, DC cables should be sheathed against unfavorable weather conditions, screened against lightning strikes and mechanically protected.

The DC main cable connects the junction combiner box to the inverter.

AC cabling connects the inverter to the loads and the electricity grid. AC cables do not need to be designed for outdoor conditions but should be three-poled for single-phase inverters (or five-poled for three-phase inverters).

Important!

Higher system voltage results in **lower currents**, which means low voltage drops in cables and hence – **low cable losses**.

Higher currents require cables with larger cross section and hence – **more expensive system**.

DC cabling, compared to AC cabling, should have the following additional features:

- Double insulation
- UV-resistant and water-resistant
- Designed for operation in wide temperature range (-40°C to 120°C)
- Designed for high voltage (more than 2 kV) operation
- Easy to mount, light and flexible

- Fire-resistant and low-toxic
- Sized for low voltage drops.

Components of grid-tied systems with battery backup

Inverters in grid-tied systems with battery backup

Grid-direct battery-based inverters, similarly to conventional grid-direct inverters, are provided with an anti-islanding feature – they cannot send power to the grid during utility outage, and certainly during such outage, no AC power can be accepted from the grid to charge the battery bank. So, the anti-islanding works as the same type of alert as in grid-direct inverters. The second important type of alert, again with regards to safety, is related to battery discharge below a certain level.

Battery-based inverters can be used in systems with energy storage – either utility-interactive battery-based systems or stand-alone systems.

A battery-based inverter for a grid-direct battery-based system performs the following functions in opposite directions:

- Converting DC power into AC power for meeting the energy needs of household devices.
- Converting AC electricity from the grid into DC electricity to charge the battery.

The following situation might occur in a utility-interactive battery-based system. The battery might be fully charged, and no loads might be plugged into the system. Therefore, the whole power generated by the PV array is sent to the grid. If the PV array, however, produces more power than the inverter could handle, the remaining power is not going to be used. Therefore, the inverter is going to operate with lower efficiency.

Grid-tied battery-backup inverters are more sophisticated and more expensive than grid-tied battery-less inverters, since apart from sending power to the grid they, are also expected to:

- Charge the battery bank after outage, and
- Provide power to all the backed-up loads during an outage.

Certainly, such inverters should have features typical for grid-tied battery-less inverters and stand-alone inverters.

Important!

Features of grid-tied battery-backup inverters:

- Battery charging – the inverter converts AC voltage (coming from either the grid or a backup generator) into DC voltage suitable for charging the battery bank
- Proper alert when the battery bank voltage is low
- Switching on a backup generator during power outages, if the batteries are at low level. When the batteries are fully charged, the inverter turns the generator off.
- High current provision for backed-up loads, including surge loads (all the loads with motors needing high starting current).

Specifications of grid-tied battery-backup inverters:

- Rated input power – compared to grid-tied battery-less inverters, here the inverter should not only be able to handle the DC power delivered by the PV array, but also to manage all the backed-up loads operating simultaneously.

- DC input voltage accepted from the battery bank – voltage values are standard, the most typical are 12V, 24V and 48V.
- Output voltage – usually 120 VAC or 240 VAC for most residential buildings.
- Output frequency – 50Hz in Europe, 60 Hz in the USA.
- Surge capacity – allows an inverter to supply much more output power than its rated value within a short period, to provide high starting current to motors (in refrigerators, water pumps, etc.)

In a stand-alone system, when selecting an inverter, the power output of the PV array doesn't need to be considered, since the battery bank is placed between the inverter and the PV array (for details, see *Inverters for stand-alone systems*).

When selecting an inverter for a grid-tied system with battery backup, however, both the PV array power output and the battery power output should be considered.

If the batteries are fully charged, and there are no loads plugged, the inverter must be able to send all the PV power into the utility grid. In such a situation, if the PV array can provide more power than the inverter is capable of processing into AC power and send into the grid, such additional power will not be used, and the PV array will not be able to operate at its maximum efficiency.

As a result, the client will send less energy into the grid.

Anything valid about PV modules and arrays (including

PV array sizing) in grid-tied systems without battery backup, is valid also for grid-tied systems with battery backup.

Inverter-chargers

If you have a system with battery backup, you could also use an inverter-charger:

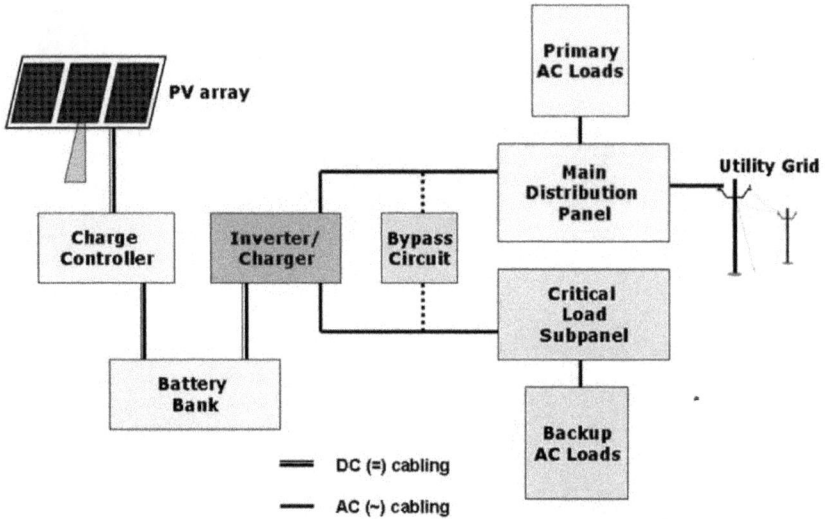

Important!

An inverter-charger:

- Converts DC power from the battery to AC power.
- Manages the battery charge from the grid via an integrated charger.
- Exports the surplus of energy to the grid. Therefore, it must have an anti-islanding protection.

Batteries

General information

Batteries are devices capable of producing and storing DC electricity.

In photovoltaic systems batteries are used to replace the photovoltaic generator:

- At night,
- During cloudy weather, or
- While the PV array is disconnected for repair and maintenance.

A battery cell is a container usually filled with a diluted acid (fluid) used as an electrolyte, with two plates (of positive and negative polarity) immersed into it. Such a battery cell is called a 'wet cell'. There are also 'dry cells' which do not contain a liquid electrolyte.

Battery cells connected together form a battery. Batteries connected together form a battery bank.

Battery features and parameters

The ability of a battery to store DC electricity is called 'capacity' and is measured in amperes-hours (Amps-hours, Ah).

Battery capacity shows how much current can be provided for a certain number of hours.

A battery of 100 Ah of capacity can provide either 100 A within 1 hour, or 10 A within 10 hours.

The optimal charge/discharge current for a battery of 100 Ah of capacity should be 10 A. Such battery can deliver this charge/discharge current for a period of 10 hours.

An important factor affecting capacity is storage temperature. The lower the temperature, the less capacity provided.

After DC electricity is stored in a battery, it can later be rendered as DC **voltage**.

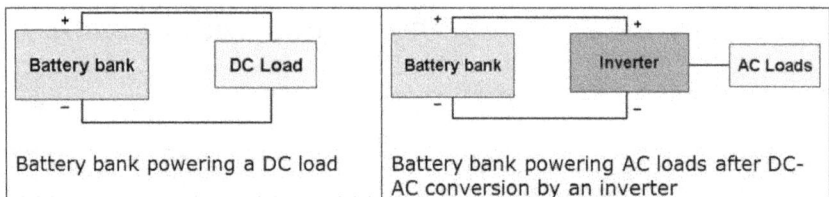

Battery bank powering a DC load	Battery bank powering AC loads after DC–AC conversion by an inverter

Standard values of battery voltage are 6V, 12V, 24V and 48V.

The higher the capacity, the longer the period of delivering the rated voltage.

After some time a battery is no longer able to deliver the rated voltage. Such a battery is said to be in a state of 'discharge'. To be capable of delivering the stated voltage, the battery needs to be 'recharged'.

Important!

The shorter the discharge/charge period, the shorter the battery life.

As a rule of thumb, for optimal battery life, the charge/discharge current should not exceed 1/10 of its capacity.

For example, if you have a battery of capacity 55 Ah, the charging current should not exceed 5.5 Amps.

In a PV system the battery bank is recharged by the PV array:

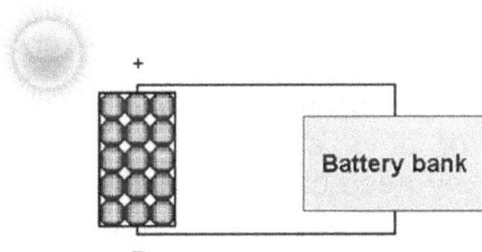

Batteries used in PV systems are rechargeable. Every battery has a certain life duration, which means that it cannot be discharged and recharged for a limitless number of times.

Important!

Vehicle batteries are not designed for frequent and deep discharge, as is the case with photovoltaics.

Therefore, vehicle batteries are not recommended for use in PV systems.

87

Important!

Other main battery parameters are:

- **Depth of Discharge** (DoD) - the extent down to which the capacity can be reduced during discharging. The lower the DoD, the longer the battery life. A battery regularly discharged down to 80% of its capacity, will have a shorter life than a battery regularly discharged to 50% of its capacity.

For deep-cycle batteries, DoD is about 80%, and for liquid electrolyte batteries, DoD is 50%.
The most practical value to use when sizing a lead-acid battery (please, read on) bank is 50% DoD.

- **Days of autonomy** (DoA, holdover) – the number of days a battery can support a load, without any need to be recharged by the PV array.

The more the days of autonomy, the more expensive the battery. DoA depends on whether a solar panel system will use a generator or not.
The typical value of DoA is **2-3 days** for hybrid off-grid systems and **3-7 days** for stand-alone (photovoltaic-only) systems.
End of battery life – it is a common understanding that the end of battery life is reached when the battery is capable of maintaining about 70-80% of its original capacity.

Specific gravity and freezing point of electrolyte

Specific gravity denotes the percentage of the acid in the battery's electrolyte. Specific gravity is measured by a hydrometer and indicates the weight of the electrolyte, compared to an equal amount of water.

The greater the state of charge of a battery, the higher the specific gravity of the electrolyte. Hence, the voltage per cell incrases, and the total battery voltage also increases. Therefore, by measuring the specific gravity of a battery upon its discharge, you will be informed about the battery's state of charge.

Freezing point

Lead-acid batteries can quickly freeze at low temperatures because the predominant part of their electrolyte is water. The good news is that the sulfuric acid in the cells acts as an antifreeze agent. Hence, the higher the percentage of aid in the water, the lower the freezing temperature.

At extremely low temperatures, however, even a fully charged lead-acid battery is likely to freeze. The typical freeze temperature of a lead-acid battery is -10°F or -23°C. When the state of charge goes down, the specific gravity goes down too. Therefore, the acid is becoming 'weaker' and 'lighter'.

State of charge	Specific gravity	Voltage per cell, volts	Voltage of 12V (6 cell) battery	Feezing point (°C / °F)
Fully charged	1.265	2.12	12.70	-57 / -71
75% charged	1.225	2.10	12.60	-37 / -35
50% charged	1.190	2.08	12.45	-23 / -10
25% charged	1.155	2.03	12.20	-16 / +3
Fully discharged	1.120	1.95	11.70	-8 / +17

89

Specific gravity, voltage and freezing point of lead-acid batteries according to the state of charge

It can be seen from the table that a fully discharged battery should be kept at temperatures above 20 °F or -7°C. If you cannot keep a battery in a warmer place, you should maintain it at a high enough charge to prevent the electrolyte freezing. The good news is that such a task can be performed by a charge controller capable of disconnecting the load when the battery voltage falls below a certain level. Certainly, this is not applicable in case of critical loads that cannot be easily turned off.

Changing the temperature of the electrolyte results in changing the charging characteristics of lead-acid batteries as well. A cold battery will accept a relatively low state of charge. A warmer battery will result in higher charge rates. In case of climate with rapidly changing temperatures (high thermal amplitudes), it is a good plan to adjust the specific gravity of the electrolyte according to the particular temperature. Thus the life cycle of the battery will be extended despite the extreme climatic conditions.

What is overcharging and overdischarging?

Important!

When used in solar panel systems, batteries must be prevented from *overcharging* and *overdischarging*.

Overcharging could lead either to a hazardous condition or to a shortening of the battery's life. The same is valid for overdischarging.

In a solar electric system, the battery bank is prevented from overdischarging and overcharging by the charge controller.

Sources:

1. Sandia National Laboratories. 1991. Maintenance and Operation of Stand-Alone Photovoltaic Systems.
2. Pop MSE, Lacho, Dimi Avram MSE. 2015. The Ultimate Solar Power Design Guide: Less Theory More Practice, Kindle Edition. Digital Publishing Ltd.

What kinds of batteries are used in solar systems?

The following types of batteries are used in stand-alone PV systems:

- Lead-acid batteries – either flooded or sealed (VRLA)
- Alkaline batteries – always sealed: nickel-cadmium (for small portable appliances or daily loads) and nickel-iron.

Important!

Lead-acid batteries are the best choice for residential PV systems.

Lead-acid batteries comprise multiple individual cells of nominal voltage 2V each.

Lead-acid batteries contain so-called 'wet cells', since their electrolyte is in a liquid state.

Every battery has its lifecycle and can only be discharged and recharged for a limited number of times.

Lead-acid batteries have the following benefits:

- Low cost
- Robust design
- High depth of discharge.

A fully-charged lead-acid battery cell has a voltage of about 2.12 V. The nominal value is accepted to be 2 V.

Lead-acid batteries are quite vulnerable to self-discharge. Regardless of the ambient temperature, when not being used, a lead-acid battery can lose up to 5% per month of its capacity. The higher the temperature, the faster the self-discharge even for a fully-charged battery.

Lead-acid batteries can be subdivided into two types:

- Flooded lead-acid batteries, and
- Valve-regulated lead-acid batteries.

Flooded lead-acid batteries have removable caps and require regular maintaining activities – checking the level of electrolyte and adding distilled water when the level is below the required minimum. The caps should be kept opened during the process of recharging to free the generated hydrogen gas.

Valve-regulated lead-acid (VRLA) batteries are also known as 'sealed lead-acid batteries'. During recharging the pressure of the hydrogen generated in the cells pushes the valves up and lets the gas free.

VRLA batteries have the following benefits:

- Require no maintenance
- Produce less hydrogen gas during recharging
- Suitable for locating in limited room space.

VRLA batteries are not widely used due to the following drawbacks:

- Greater cost
- Shorter life span
- Required charging voltage should not be exceeded. Otherwise too much hydrogen gas might be generated, that cannot be fully released by the valves.

Non lead-acid batteries

Use of less antimony and more calcium, cadmium or strontium leads to less gasing and lower water consumption.

Such batteries, however, should not be discharged more than 20%, since their life cycle shortens quickly.

Self-discharge is not a big issue in sealed lead-acid batteries, as they normally have hybrid lead-calcium or lead-antimony electrodes. Self-discharge for such batteries can be minimized by keeping them in a cool place at a temperature at 10÷15°C (50÷59°F).

Nickel-cadmium (Ni-Cad) batteries are similar to lead-acid batteries regarding metal plates and liquid electrolyte. Instead of lead, however, the plates are made of nickel hydroxide (positive ones) and cadmium oxide (negative ones). The electrolyte is potassium hydroxide rather than sulfuric acid.

The cell voltage of a typical Ni-Cad battery cell is 1.2 volts, unlike the 2V voltage per lead-acid cell.

Ni-Cad batteries are less vulnerable to freezing and high temperatures than lead-acid batteries.

The self-discharge of Ni-Cad batteries is similar to the self-discharge of lead-acid batteries and varies within 3-5% per month. A Ni-Cad battery can be fully discharged without damage. Moreover, there is no sulfation in Ni-Cad batteries.

A sheer drawback of nickel-cadmium batteries is their high initial cost. This, however, is easily compensated by their low maintenance cost and the longer life cycle because they are less affected by overcharging and overdischarging. The lack of intense maintenance activities and low maintenance costs makes Ni-Cad batteries ideal for cases where batteries are located in a hard-to-access and/or dangerous place.

Gelled or **Absorbed Glass Mat (AGM) batteries**

are known as sealed batteries. They are also known as 'captive electrolyte batteries'. These batteries have hybrid lead-calcium or lead-antimony electrodes. In case of captive electrolyte, you don't have to charge the battery high enough to let gasses to be released during the process of charging. Since the electrolyte is not a liquid, a captive electrolyte battery can be used in any position, even upside down, being able to deliver its full capacity. These batteries are shallow-cycle and discharging them more than 20% of the capacity reduces their lifecycle. Therefore, Gelled and AGM batteries are not the best fit for solar electric systems. Despite their low self-discharge, these batteries should be avoided to use below -20°C (-4°F) and above +50°C (+122°F).

Sealed (gelled) batteries are suitable for **mobile** applications of **small, low-current** stand-alone photovoltaic systems.

The benefits of sealed batteries:
- Easy for handling and transportation
- Safety (no acid, no gassing during charging)
- Maintenance-free (no need to add distilled water).

The drawbacks of sealed batteries:
- Should be replaced every few years
- Overcharging reduces battery capacity
- Relatively high cost: far from unaffordable but anyway much higher than flooded batteries.

Important!

Sealed (gelled) batteries are usually preferred in mobile stand-alone solar electric systems.
For non-mobile stand-alone solar electric systems, lead-acid flooded batteries are recommended due to their:

- Long life
- High amount of delivered energy
- High reliability
- High performance (higher than gel batteries)
- Low cost.

Flooded batteries, however, require serious attention, regular maintenance activities and high commitment by the user, including a special room for placing the battery bank.

Comparison between the main battery types:

	Lead acid, unsealed, flooded, deep cycle battery	Lead acid, sealed, flooded, shallow cycle battery	AGM battery	Ni-Cad battery
DoD	40-80%	15-25%	15-25%	100%
Self discharge rate, %/month	5	1-4	2-3	3-6
Typical capacity, Amps-hrs/ft³	1000	700	250	500
Typical capacity, Amps-hrs/lb.	5.5	4.6	2.2	5.0
Min. environment temperature	+20 F (-6 C)	+20 F (-6 C)	0 F (-18 C)	-50 F (-46 C)

Source:

1. Sandia National Laboratories. 1991. Maintenance and Operation of Stand-Alone Photovoltaic Systems.

Connecting batteries

1) Series connection

For batteries connected in series, the total voltage is a sum of the individual battery voltages, while the total capacity is the rated capacity of a single battery:

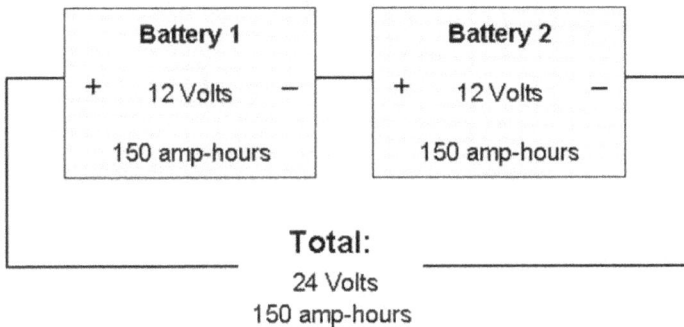

You should, however, remember the following:

Important!

If you connect batteries of **different capacity** in series, the total capacity obtained is equal to **the lowest capacity** in the string.

Here is an example:

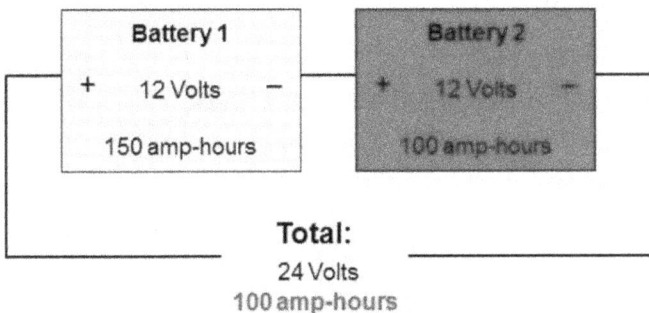

Important!

Avoid connecting in series a set of batteries of different capacities!

2) Parallel connection

For batteries connected in parallel, the total capacity is a sum of the individual capacities of the connected batteries, while the total voltage is the rated voltage of a single battery:

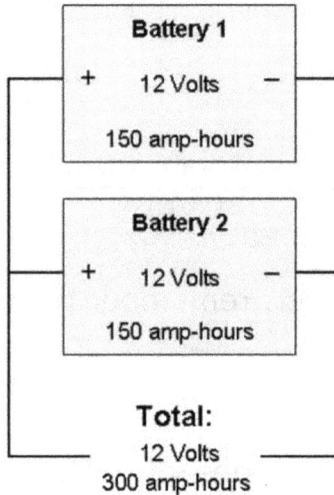

```
┌──────────────────────────┐
│        Battery 1         │
│  +      12 Volts      −   │
│                          │
│      150 amp-hours       │
└──────────────────────────┘

┌──────────────────────────┐
│        Battery 2         │
│  +      12 Volts      −   │
│                          │
│      150 amp-hours       │
└──────────────────────────┘

          Total:
         12 Volts
       300 amp-hours
```

Important!

Recommendation: Connect identical batteries in series!

If you need a higher capacity, you should connect maximum two strings in parallel.

Otherwise, the defective battery in a parallel string will start to act as a load to the adjacent 'good' string connected in parallel. Thus, the overall battery bank capacity will be reduced.

The picture below presents a battery bank comprising a mixed connection of batteries to increase both overall voltage and capacity:

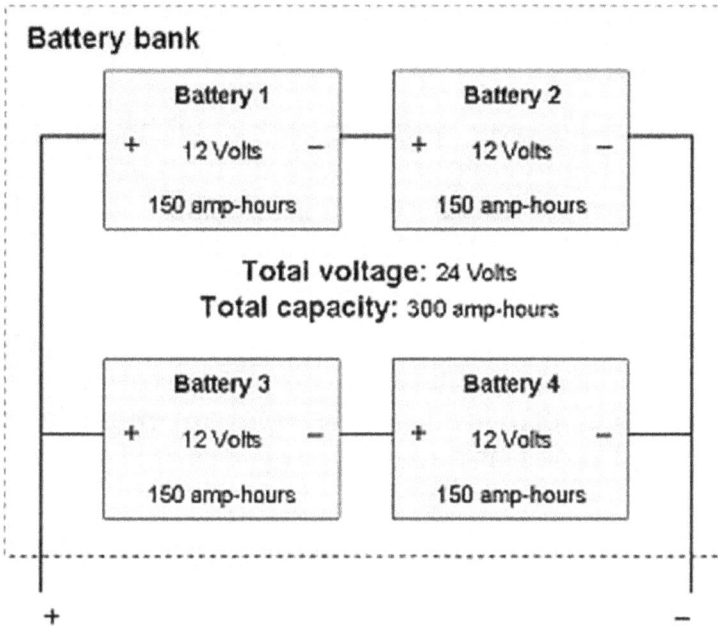

Source:

Pop MSE, Lacho, Dimi Avram MSE. 2015. The Ultimate Solar Power Design Guide: Less Theory More Practice, Kindle Edition. Digital Publishing Ltd.

Battery safety

If not handled, installed or maintained correctly, batteries can be a hazardous component of any solar electric system.

Potential risks can include work with dangerous chemicals, heavyweight, as well as high voltages and currents, which could result in possible explosions, burns, corrosive damages, injuries and electric shocks to users.

Therefore every user should follow the instructions for handling, installation, operation and maintenance, provided by the relevant manufacturer to avoid all types of hazards.

Important!

General battery safety rules:

- Always draw a battery diagram before connecting the batteries to a battery bank.
- Make sure the battery room is properly ventilated. Never use a living room for storing batteries.
- Wear protective clothing (especially eye protection) while handling batteries.
- Use proper tools while connecting battery cells.
- Always have fresh water accessible in case that battery electrolyte gets in touch with your skin and eyes. In the event of an accident, flush with fresh water for 5 to 10 minutes and do not hesitate to contact a physician afterward.
- Keep open flames away from batteries and don't allow smoking nearby.

- Before working with a battery bank, disconnect any source of charging and discharging that might have been connected to it.
- Never lift a battery by holding by the terminals. Always lift a battery by holding it from the bottom.
- Discharge the static electricity of your body before touching the terminals.
- Always use cables of same length for battery connection.
- Keep number of parallel battery connections to a minimum.
- Connect batteries **last** among all components.
- Regularly clean corrosive film formed on the terminal posts.
- Don't try to check the amps across the battery terminals.

Components of stand-alone (off-grid) systems

Batteries for stand-alone systems

Since in most stand-alone PV systems, not all the generated electricity is consumed right away, an energy storage (batteries) is required.

Important!

Batteries are the heart of every stand-alone PV system.

Batteries have three important features:

- Autonomy (the main advantage of stand-alone systems) – through proper selection of battery parameters (capacity and depth of discharge; refer to chapter 'Electricity Basics') the PV system power output is matched to energy needs of a household, without any dependence on the utility grid.
- Provision of a stable DC voltage enabling the inverter or the loads to work properly and not to get damaged.
- Ability to provide higher currents than PV array. Such high currents are needed upon starting motors when surges (abrupt and high increases in current, known as 'spikes') are likely to occur.

The battery is known as the heart of any stand-alone photovoltaic system. It is the primary component that makes off-grid systems different than grid-tied systems.

The cost of batteries alone is between 25% and

50% of the total cost of a stand-alone system.

Batteries of longer lifecycle cost more, but they are cheaper to maintain.

Car batteries are not suitable for PV systems, since they are not designed for deep discharges.

In stand-alone PV systems, the following types of batteries are used:

- Lead-acid batteries – flooded; sealed (VRLA)
- Alkaline batteries – nickel-cadmium (for small portable appliances or daily loads); nickel-iron
- Lithium-Ion batteries – used in electrical vehicles and small airplanes, cell phones and other portable electronic devices; currently they are trying to find their place also in medium-scale and large-scale solar power systems.

Lead-acid batteries

Important!

Lead-acid batteries are **the best choice** for residential PV systems. They offer you '**the best value for the price**'.

Lead-acid batteries are subdivided into two types:

- Flooded lead-acid batteries
- Valve-regulated lead-acid batteries.

For non-mobile stand-alone systems, lead-acid (flooded) batteries are recommended, due to their:

- Long life
- High amount of delivered energy
- High reliability
- High depth of discharge

- High performance (better than gel batteries)
- Low cost.

Flooded batteries do have their drawbacks, since they require:

- A separate, safe and well-ventilated room for installation
- Regular maintenance activities
- Serious attention and high commitment by the user.

Valve-regulated lead-acid (VRLA) batteries are also known as 'sealed lead-acid batteries'. During recharging the pressure of the hydrogen generated in the cells pushes the valves up and lets the gas free.

VRLA batteries have the following advantages:

- Do not require maintenance by the user regarding regular level checking and adding water.
- Produce less hydrogen gas during the recharge process.
- Are suitable for locating in a limited room space.

VRLA batteries, however, are not widely used and still are not preferred to flood lead-acid batteries, due to the following reasons:

- More expensive
- Shorter life span
- The charging voltage should not be exceeded.

For grid-tied systems with battery backup **Absorbed Glass Mat (AGM) VRLA batteries** are recommended.

An AGM battery is a sealed lead-acid battery that is completely maintenance-free. The battery has its electrolyte absorbed in plates made of a sponge-like mass of matted glass fibers. Electrolyte in such a form is non-spillable, therefore AGM batteries are flexible to various types of positioning.

Alkaline batteries

Gel (sealed) batteries are suitable for **mobile** applications of **small, low-current** stand-alone photovoltaic systems.

The advantages of gel batteries are:

- Easy for handling and transportation
- Safety (no acid, no gassing during charging)
- Maintenance-free (no distilled water to add).

Their disadvantages are:

- They have to be replaced every few years
- Overcharging reduces battery capacity
- High cost – far from unaffordable but anyway much higher than the cost of flooded batteries.

Nickel-cadmium batteries are an excellent choice for mobile applications, since they:

- Have great number of discharge cycles
- Are less affected by temperature changes than lead-acid batteries
- Are less dangerous
- Are maintenance-free
- Are easy for transportation.

For residential applications, however, nickel-cadmium batteries are not advantageous, due to their high price and tough recycling. But if you can afford a nickel-cadmium battery for your home, it is the best choice.

RV (Recreational Vehicle) or Marine batteries are another battery type typically used in small systems, such as boats or campers. Although being deep-cycle batteries, they differ by typical deep-cycle batteries, as follows:

- They are smaller in size and are less expensive.
- They are usually either of AGM- or gel type (i.e. they are maintenance free).
- Their voltage is 12V or 6V – to get a total of 12 V, you connect two batteries in series – as is the case with golf cart batteries).
- Withstand to (3 to 4 times) less charge/discharge cycles than residential deep-cycle batteries.
- Can be used more successfully to start an engine although they show a lower performance than conventional vehicle 'cranking' batteries.

RV/Marine batteries are actually a compromise between vehicle batteries (also known as 'cranking' batteries used for engine starting) and deep-cycle batteries (typically used in residential solar electric systems). RV/Marine batteries are used in small photovoltaic systems where the size of the battery is often as important as its performance.

Lithium-Ion Batteries

Lithium-Ion (Li-Ion) batteries are quite suitable for small solar electric systems, especially portable and handheld devices. Over the years the Li-Ion technology has improved and has become a reasonable alternative to lead-acid batteries in small-scale photovoltaic systems and are becoming more popular also for larger solar electric systems.

Advantages of Li-Ion batteries

- Low weight – about 30% of the size of a lead-acid battery of the same capacity.
- Lower space for housing – about 50% of the space needed for housing a lead-acid battery of the same capacity.

- Higher recommend charging current for ensuring a maximum battery life – about 30% of battery capacity compared to recommended 10% of the battery capacity for lead-acid batteries.
- Higher maximum discharging current.
- Recommended maximum discharge depth between (70-80)%, compared to 50% for lead-acid batteries.
- Four times (on average) higher battery cycle life.
- Twice higher battery life under recommended usage conditions, which translates into about 10 years for Li-Ion batteries, compared to 5 years for lead-acid batteries.
- Perform better in hot climates.
- Fully maintenance-free.

Disadvantages of Li-Ion batteries

- Higher price
- A battery management system (BMS) is a must. Such a BMS is used both to manage and to protect the battery from a thermal runaway event upon which there is a risk both of fire and chemical contamination. The BMS also disconnects the Li-Ion battery from the loads upon any abnormal variations with temperature, current or voltage.
- The battery voltage operating window might be incompatible with the input operating window of the inverter, since most inverters are designed to operate with lead-acid batteries. This might lead to problems in the communication with the BMS.

Although for Li-Ion batteries the higher price could be justified by the higher lifecycle, the requirement for a battery management system (BMS) is a major obstacle for using these batteries, especially in large solar electric systems.

Important!

Li-Ion batteries in summary:

- Today the application of Li-Ion batteries for solar power is viable in cases where lack of space and low battery bank weight are more important than battery price.
- Although being a higher initial investment, lithium-based batteries, compared to lead-acid batteries, could turn out to be a lower cost solution in the long run for solar power systems that require a higher depth of discharge and more frequent charging/discharging.

Battery sizing example

The primary task in battery sizing is estimating battery capacity you need for your solar electric system.

Battery capacity required =
= (Average daily energy target x Days of Autonomy x Battery Temperature multiplier) ÷ (Depth of Discharge x Cable Losses Factor x System voltage)

[Pop MSE, Lacho (2015-10-26).The Truth About Solar Panels: The Book That Solar Manufacturers, Vendors, Installers And DIY Scammers Don't Want You To Read, Kindle Edition. Digital Publishing Ltd]

Therefore, to calculate the required capacity of the battery bank, we need to know the following:

- The average daily energy target – see the example below how to estimate it
- The Days of Autonomy – we assume a value of 3
- The Battery Temperature Multiplier:

Ambient Temperature	Battery Temperature Multiplier
80°F / 26.7°C	1.00
70°F / 21.2°C	1.04
60°F / 15.6°C	1.11
50°F / 10.0°C	1.19
40°F / 4.4°C	1.30
30°F / -1.1°C	1.40
20°F / -6.7°C	1.59

- The Depth of Discharge (DoD) – we assume a value of 0.5;
- The cable losses factor – it is a product of two multiples – AC cable losses factor and DC cable losses factor. If each of them is assumed 0.975, the total product is 0.95.
- The system voltage – we choose 24V in order to use cables of smaller cross-section, that is, less expensive ones.

For residential off-grid solar electric systems, the system voltage is usually selected either 12V or 24V (48V is less common).

There are several criteria for selecting system voltage:

- The length of cables between the solar modules and the battery. Higher voltage (24V instead of 12V) means lower current, so cables of smaller cross sections should be used. Furthermore, DC devices of higher power rating consume higher current – for example, a 12V DC device of power rating of 400 W consumes a current of 33A, while a DC device with the same rated power, but for designed for 24V, consumes a current of 16.7A.
- The system size – low power systems usually operate at 12 V
- Whether the system will contain a high power inverter. Inverters of power rating of up to 2,000W are rated for 24V, while inverters with higher rating are rated for 48V.
- Availability of DC loads – some DC loads only support 12 VDC.

Here is an example of how to estimate the average daily electricity target for a household.

AC loads table

Loads	Rating, W	Qty	Total power, W	Hours of use/day	Days of use/week	7 days/ week	Avg. daily use, Wh
CF lights	15	4	60	5	7	7	300
Incandesc. bulbs	75	2	150	0.3	7	7	50
TV	100	1	100	4	7	7	400
Laptop	50	1	50	8	5	7	286
Microwave	800	1	800	0.2	7	7	160
Fridge	100	1	100	24	7	7	450
Clothes washer	1200	1	1200	1	4	7	686
						Total daily use, Wh	1,646

Total power = Rating x Qty

Average daily use = (Total power x Hours of use per day x Days of use per week) ÷ 7 days per week

The customer does not intend to use any DC loads.

We assume: inverter efficiency 0.92, batteries depth of discharge 50%, cable loss factor 0.95 (i.e. cable losses 5%) and system voltage 24V.

Average daily energy target =
(Average AC daily load ÷ Inverter efficiency) + Average DC daily load = (1,646 ÷ 0.92) + 0 = 1,789 Wh = 1.789 kWh

Certainly, if the off-grid system is DC-only, inverter efficiency is assumed 1.

So, according to the formula:

Battery capacity required =
= (Average daily energy target x Days of Autonomy x Battery Temperature multiplier) ÷ (Depth of Discharge x Cable losses factor x System voltage) =
= (1,789 x 3 x 1.04) ÷ (0.5 x 0.95 x 24) = 490 Ah

After we have calculated the required capacity of the battery bank, now it is time to select a particular battery model.

Considering the value of 490 Ah, we can select two 24V-batteries, each of a capacity of 250 Ah, which can be connected in parallel:

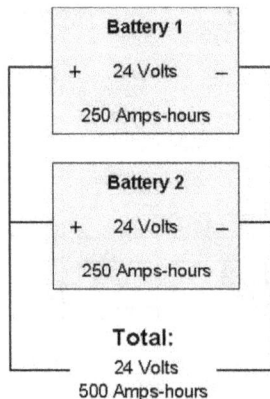

Battery 1
+   24 Volts   −
250 Amps-hours

Battery 2
+   24 Volts   −
250 Amps-hours

Total:
24 Volts
500 Amps-hours

Source:

1. Pop MSE, Lacho, Dimi Avram MSE. 2015. The Ultimate Solar Power Design Guide: Less Theory More Practice, Kindle Edition. Digital Publishing Ltd.

Batteries in grid-tied systems with battery backup

In a grid-tied system with battery backup, the battery bank has the following distinctive features:

- Unlike in stand-alone systems, in a grid-tied system with battery backup, batteries are fully charged and expecting possible power outage during most of the time.
- The loads served by the battery bank during a power outage should be the essential ones, like some lights, a refrigerator, a PC/laptop, and a water pump, for example. The load balance, therefore, should be performed for this minimum amount of loads that might also be called 'critical loads'.
- The capacity of the battery bank is never estimated to cover **all** the energy needs of the building, but rather the essential needs during an outage. The battery capacity should be therefore estimated to cover the duration of the outage, which means that the PV array is not used as a regular daily battery charger, as is the case with stand-alone systems.
- For stand-alone systems, usually 3 days of autonomy (holdover) are chosen as a minimum when sizing the battery bank. For grid-tied systems with battery backup the maximum holdover is one day; power outages of 24 hours are unlikely to happen in general.

Blocking diodes

What is the role of a blocking diode? First, let's recall what a diode is.

A diode is an electronic element allowing the current to flow only in one direction. In the opposite direction, it serves as an open switch, thus preventing the current from passing through the circuit, if the current direction changes for whatever reason. In other words, the resistance of a diode is very low in its conductivity direction and very high in the opposite ('reverse') direction.

Commonly a diode is made of silicon. When we use a diode, it is critical to pay attention to whether the direction of the current we want to let through the circuit coincides with the direction of the diode's conductivity.

The role of the blocking diode is to prevent the battery bank from discharging through solar panels during the night or when solar panels are shaded. Blocking diodes are placed in series in regards to a solar panel or solar array.

While the sun is shining, the solar array is charging the battery bank and the current flows via the blocking diode from the solar panels to the battery:

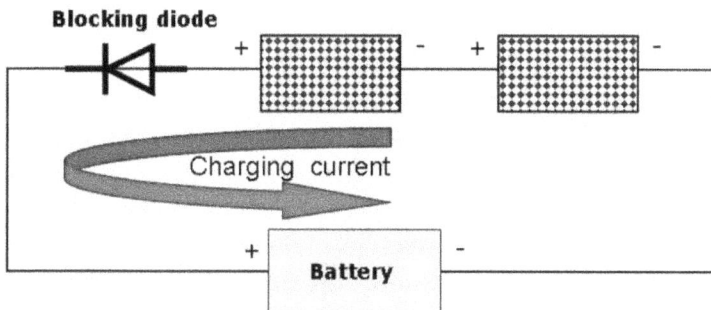

115

If no blocking diode were used, the battery bank would start discharging via the solar panels during the night or when the solar array gets shaded. The direction of the discharge current would be from the battery bank to the solar panels.

However, if we put a blocking diode between the battery and the solar panels, with conductivity direction as depicted in the picture, the diode will prevent discharge current from flowing via the solar panels.

Important!

If a solar array includes blocking diodes, here is how to check whether they are operating correctly:

- The positive side of the diode should be on the array side.
- Measure the voltage drop across each diode upon current flowing from the array to the batteries. If the measured voltage drop is between 0.5 and 1.0 volts, the diode is operating correctly. If the voltage drop is higher, the diode is defective and must be replaced.

Charge controllers

The most important function of a charge controller is to prevent a battery bank from overcharging and overdischarging.

Charge controllers (also known as 'charge regulators' or 'battery chargers') are used for proper maintenance of the battery bank.

One of the most common problems of batteries is that they cannot be discharged excessively or recharged too often. A charge controller controls the charge by managing the battery voltage and current properly.

Charge controllers are intended to protect the battery and to make it live as longer as possible while keeping the PV system efficiency.

Important!

The primary functions of charge controllers are:

- Protecting the battery from overcharging by limiting the charging voltage.
- Protecting the battery from deep and/or unwanted discharging. The charge controller automatically disconnects the loads from the battery, when the battery voltage falls below a certain Depth of Discharge value.
- Preventing the reverse current through PV modules at night.
- Providing information about battery state of charge.

Other essential functions of charge controllers are:

- Switching between various recharging modes depending on the particular battery type.
- Protection against overloading and short circuit.
- Temperature compensation – application of individual recharging mode by voltage according to the battery temperature.
- Protection against overvoltage.
- Various indicator lights – battery charging level, charging current, charging voltage, time remaining till the fully charged state, etc.
- Further charging of the liquid electrolyte cells.

For higher currents, more than one controller is used. In such a case the PV array should be divided into sub-arrays and each sub-array should be connected to its controller. All of them however, can use the same battery bank.

Here is how charge controllers prevent overdischarging:

- Lights or buzzers for low battery voltage are activated.
- Certain loads are turned off at a preset level of state-of-charge.
- A backup power generator is turned on.

Loads are turned off to prevent a battery bank discharge. Such a feature is called 'load management' and is implemented via a low voltage disconnect (LVD) mechanism. Load management by LVD makes all DC loads automatically shut down. Therefore, to prevent undesirable disconnect of certain DC loads, such loads are to be connected directly to the battery. Certainly, in such a case battery overdicharge is still likely to occur due to these loads.

Lights and/or buzzers could also be used to alert low battery voltage and therefore disconnecting certain critical loads. However, the user could ignore the alert, so there is still a risk of overdischarging and reducing battery life.

Important!

Charge controllers control DC loads only.
AC loads are to be controlled by an inverter.

Charge controller types

The main charge controller types available today are PWM (Pulse Width Modulation) and MPPT (Maximum Power Point Tracking) ones.

- PWM controllers are less expensive than MPPT ones. A PWM controller prevents batteries from overcharging and overdischarging in a way that extends the life of batteries slightly more than an MPPT charge controller does, at the expense of lower (20% on average) efficiency compared to MPPT controllers.

A PWM charge controller achieves such extension of battery life as a result of applying advanced switching algorithm and pulse width modulation combined with pulse charging. Pulse charging prevents lead plates of the battery from sulfation.

- MPPT controllers is more expensive than PWM ones. An MPPT controller prevents batteries from overcharging and overdischarging while transferring more power (up to 20% more) from solar modules to either the battery bank or the system loads. It does not extend the life of batteries as efficiently as PWM charge controller does.

Some MPPT charge controllers are provided with a 'voltage step-down' feature. The voltage step-down feature allows a PV array of a higher voltage to be connected to a lower voltage battery bank – for example, a 48 V array to a 24 V battery bank. Without a step-down feature, the total voltage of the array must be equal to the battery bank voltage.

The step-down feature provides the following benefits:

- Higher PV array voltages allow the use of cables with smaller cross sections between the array and the controller thus reducing wiring costs. Therefore, more PV modules can be connected in series and fewer modules in parallel.
- Opportunity for PV array expansion without any need to increase the size of wiring.

How to select the right type of charge controller?

What kind of charge controller to choose depends on the particular case and is a tradeoff between getting more power from solar panels and extending the battery life.

Solar modules have a life cycle of up to 20-25 years. Battery life is roughly estimated between 1 and 5 years, depending on battery type and depth of discharge. On average 2 years of battery life could be assumed.

Important!

If your system is more prone to deep discharging and overcharging, and therefore the battery life appears as a primary concern, you may opt for a cheaper PWM charge controller at the expense of the reduced power provided by the system. Such production losses can be compensated by increasing the number of solar modules installed.

If providing more power to the loads and the battery bank is more important than extending the battery life, you may opt for the more expensive MPPT charge controller.

PWM controllers are very suitable for small wattage solar electric systems.

MPPT controllers are always more efficient in 'squeezing' more solar power from solar modules than PWM controllers. MPPT controllers lead to roughly 20-30% higher power output in winter and 10-20% more in summer.

Important!

Do you know that an incorrectly selected PWM charge controller can result in losing 50% of the available solar power in an RV system?

An incorrect selection of a PWM charge controller is a common mistake made by RV owners.

They get a high voltage solar panel at lowest cost per Watt, connect this solar panel or these solar panels to a PWM charge controller, and subsequently lose almost 50% percent of the available solar power.

Why does that happen?

Let's consider a 220 W solar panel with:

- Maximum power point voltage Vmpp =29.1 V
- Maximum power point current Impp =7.56 A

Let's imagine this solar panel connected to a simple RV vehicle solar power system consisting of a solar panel, a charge controller and a 12V battery bank.

As you know, a PWM charge controller is sized regarding the current delivered by the solar panels. So, the PWM charge controller will provide a charging current of 7.56A to a 12V battery bank. If we neglect all the losses of the components of this solar power system, the PWM will deliver only 7.56 x 12V = 90W of power to the battery bank.

In other words, we've lost about 130W of the available solar panel's 220W power!

If we use a Maximum Power Point Tracking (MPPT) charge controller, the current provided to the battery bank will be boosted up to 220W ÷ 12V = 18.3A by such controller.

Such a boost in current is ensured by a current booster, which is an inherent part of every MPPT charge controller. So, in this case, the battery bank will receive 18.3A x 12V = 220W power that could be stored in it.

In an ideal case with no component loses, all solar panel generated power will be stored in the battery bank.

The moral of the story: To minimize power losses when employing PWM charge controller, always connect a solar panel with maximum power point voltage Vmpp voltage closer to the battery bank's voltage.

The second option is to consider the usage of an MPPT charge controller. Although being the most expensive, its high efficiency will pay off in the long run.

Important!

In hot climates, MPPT controllers tend to lose their main advantage – higher efficiency.

In hot climates, the efficiency of an MPPT controller is only 10% better than the efficiency of a PWM controller. In cold climates, an MPPT controller might increase its efficiency up to 40% more than a PWM controller.

For small systems, however, the additional cost of the MPPT controller could be money better spent on a greater number of solar modules to offset efficiency losses, in combination with a less expensive PWM controller.

Another issue worth considering is that the average warranty period is 5 years for charge controllers and 20-25 years for solar modules. Therefore, investing in larger solar arrays, along with buying a PWM charge controller, might be the better option to choose.

Source:

Pop MSE, Lacho, Dimi Avram MSE,(2016-07-27), Top 40 Costly Mistakes Solar Newbies Make: Your Smart Guide to Solar Powered Home and Business, Paperback Edition, Digital Publishing Ltd.

Charge controllers in summary

PWM charge controller

Pros:

- Prevents batteries from overcharging and overdischarging by extending the battery life slightly more efficiently than an MPPT charge controller does.
- Less expensive than MPPT charge controllers – price starts from $25 to $250, depending on the power required.

Cons:

- Not optimized to ensure optimal load to solar panels, which translates into up to 20% efficiency (on average) loss of solar production. You must add more panels to compensate such efficiency losses.
- System voltage must exactly match the solar panel nominal voltage, i.e. you have to use solar panels of 12 V nominal voltage, a charge controller of 12 V nominal voltage and a battery bank of 12 V nominal voltage.
- Might create audio or RF noise due to used Pulse Width Modulation.
- Not very suitable for scaling up your system. You can get PWM controllers rated at up to 120 Amps, however, the most common rating is up to 60A.

MPTT charge controller

Pros:

- Prevents batteries from overcharging and overdischarging.
- 'Squeezes' more power (up to 20% on average) from solar panels by ensuring optimal load to solar panels.
- Reaches maximum efficiency in cold climates (up to 40% increase in efficiency).
- The solar panels nominal voltage could be higher than system voltage, i.e. solar panels of 48 V nominal voltage connected to charge controller charging a battery bank of 12 V nominal voltage.
- You can get an MPPT controller rated at up to 200A, however, the most commons rating are up to 80 Amps.
- For MPPT controllers, the warranty is typically longer than for PWM ones.
- MPPT controllers provide a better opportunity for system growth than PWM ones.

Cons:

- More expensive – the price of a good MPPT controller starts at $500 to $x1,000.
- In hot climates, MPPT controllers lose their main advantage – the higher efficiency. In hot climates, the efficiency of an MPPT controller might decrease to only 10% (on average) compared to a PWM controller, while in cold climates solar energy squeezing of an MPPT controller might increase up to 40%, compared to a PWM one.

Is your charge controller suitable for any battery type?

Most charge controllers offer you the option to select the battery type – a lead-acid or sealed one.

Configuring the battery type is important since it enables you to match the minimum and maximum battery voltage with the rate or charging and discharging.

As Li-Ion batteries are relatively new for off-grid solar system, there are few charge controllers supporting these batteries. For the sake of protection of the solar system from any sudden disconnects of components, Li-ion batteries are typically provided with a battery management system communicating both with the charge controller and the inverter.

If for example, the battery state of charge drops below a certain level, such battery management system will provide protection to the battery by disconnecting it from the load. Such a protection, however, will prevent the battery from further charging.

Li-Ion batteries require a special battery management system (BMS) to operate. Such a battery management system is used to protect the individual battery cells. The BMS is used both to manage and to protect the battery from a thermal runaway event upon which there is a risk both of fire and chemical contamination. The BMS also disconnects the Li-Ion battery from the loads upon any abnormal variations with temperature, current or voltage, and it also measures the battery State of Charge (SoC) and State of Health (SoH).

128

The SoC is the amount of energy used compared to the energy available in a fully charged battery, while SoH measures how the capacity of a battery has deteriorated over time compared to the original capacity. Manufacturers usually recommend a battery with SoH 80% and less as unusable.

Sometimes communication between the battery management system and the charge controller can be problematic. One of the reasons is that BMS for Li-Ion are often proprietary and lack a standard communication platform.

This fact is likely to change since Li-Ion batteries are gaining popularity in medium-scale and large-scale solar electric systems.

For example, to protect the battery, the BMS will disconnect it from the load if the state of charge has dropped below a certain threshold. The charge controller might think that the battery has been removed and might stop recharging it.

What is needed here is a reliable communication line that will reconnect the charge controller to the battery thus preventing a shutdown of the whole PV system. As a Li-Ion battery needs to be well protected, such a mandatory communication line makes things some more complicated.

Source:

O'Connor, Joseph P., Off Grid Solar: A handbook for Photovoltaics with Lead-Acid or Lithium-Ion batteries. CreateSpace Independent Publishing Platform. 2016.

Do you always need a charge controller?

According to many reliable sources, if solar panels wattage is about 10% of battery capacity, you don't need a charge controller.

This claim is also based on a practical research and experience.

For example, if you have a 10W solar panel connected to a 100Ah battery bank, you may consider not using a charge controller.

The idea of such concept is that if the battery capacity is high enough, the solar panel will never overcharge the battery bank. On the other hand, a large battery capacity guarantees that the battery bank will never be fully discharged.

However, such a statement is true if the load is always connected to a particular solar system. In practice, if such a solar system serves a boat or recreational vehicle (RV), in many situations the load might be turned off for weeks, and the danger of eventual overcharging exists.

So, if you are an owner of a boat or an RV, or for whatever reason, you turn off the loads from the solar system with a high capacity bank for a very long time, you should consider using a charge controller.

Charge controller maintenance tips

Here are some important issues recommended for charge controller maintenance support:

- Install the charge controller in a clean area, where no corrosion is possible.
- Make sure to provide enough space for cooling.
- Protect the charge controller from direct sun and water flow.
- Check at the time of mounting whether all the LED are working correctly.

Inverters for stand-alone systems

The primary function of a stand-alone inverter is converting the output voltage of either the battery bank or the solar array to AC voltage.

Not every off-grid solar system needs an inverter. The inverter is not needed if power is to be provided to DC loads only:

a) Inverter-less off-grid photovoltaic system with a battery bank:

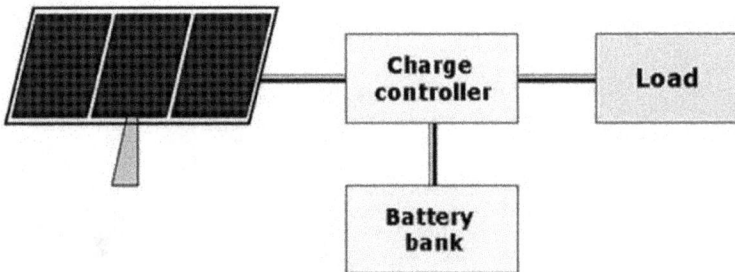

b) Inverter-less off-grid photovoltaic system without a battery bank:

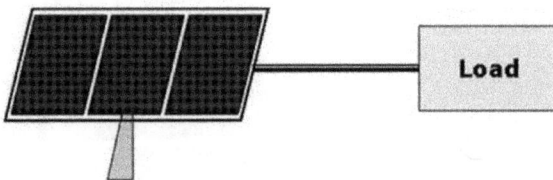

Important!

Grid-tied and off-grid photovoltaic systems use different kinds of inverters.

Since inverters for stand-alone systems are not

connected to the grid, they do not need an anti-islanding protection (refer to section 'Inverters for grid-tied systems' in the previous chapter).

Important!

The most important features of off-grid inverters are:

- Providing enough power for all the connected electrical devices.
- Generating a stable, sinusoidal AC voltage.
- Withstanding to electrical surges (also known as 'spikes') created by loads with motors.
- Low energy consumption in standby mode.
- Withstanding to tolerances in battery voltage.
- Alerting low battery capacity.
- Battery charging – converting the AC voltage coming from a backup generator (if any) to DC voltage for charging the battery bank.
- Overcurrent protection.

There are two types of inverters for off-grid systems.

The first one is the off-grid inverter directly connected to a solar array, thus providing AC power directly to AC loads, as shown below:

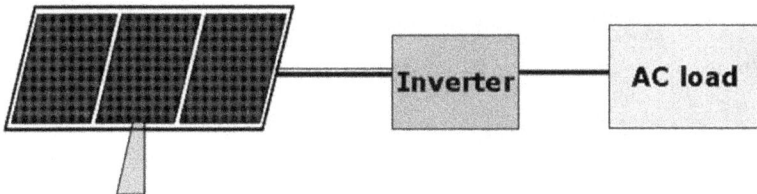

The second type is a battery back-up inverter that is connected to the battery – either directly:

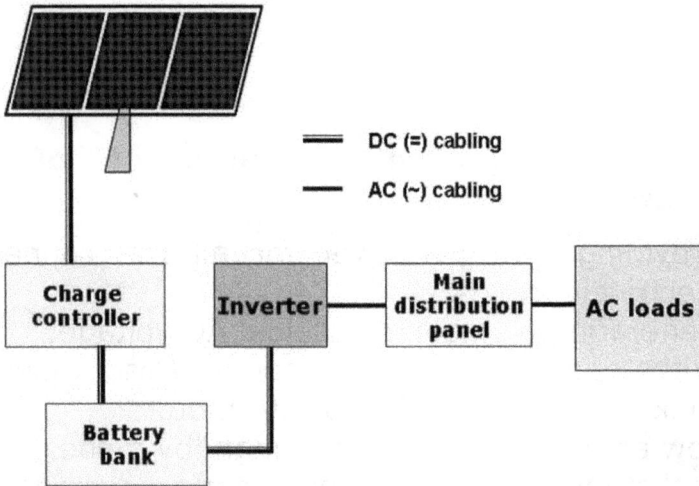

or by a DC breaker:

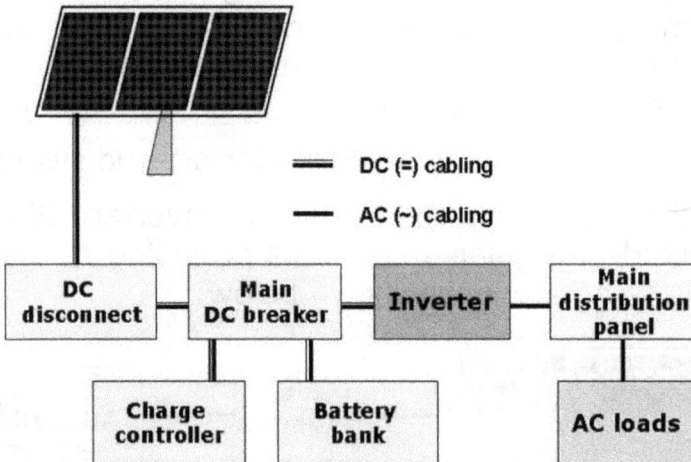

To efficiently convert battery DC power to AC power, the inverter's input voltage range must match the voltage range of the battery bank.

The voltage of the battery bank reaches the lowest value when the batteries are discharged and the highest value when the batteries are fully charged.

Furthermore, an off-grid inverter usually comes equipped with a low voltage sound alarm warning you ahead of time that the battery voltage is about to drop below the critical discharging point. Once this point is reached, the inverter starts shutting down to avoid any further discharge that is dangerous for the life of your batteries.

There are three different types of stand-alone inverters currently available on the market, with regards to produced type of voltage wave – sine-wave, quasi (modified) sine-wave and square-wave ones.

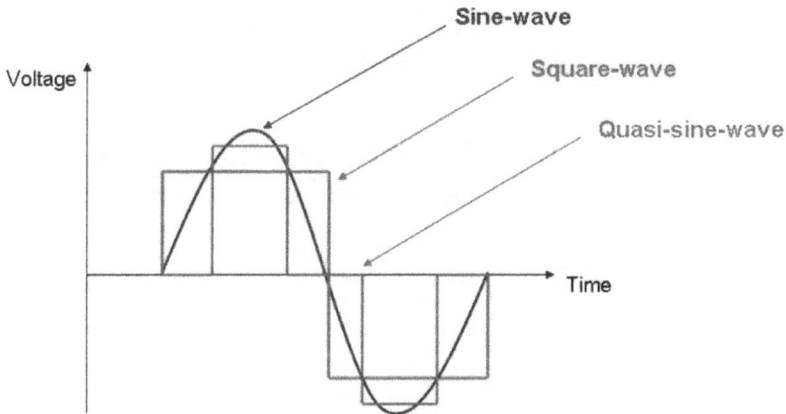

Square-wave inverters have a 'square' shape of the output current. They provide relatively weak output voltage control and significant harmonic distortion. Therefore, square-wave inverters are applicable mostly for incandescent lights and small resistive heating loads. It is not recommended to use square-wave inverters in residential solar electric systems, since most of the household equipment can easily get damaged. Square-wave inverters are the least expensive but they are also the least efficient.

135

Modified square-wave (also known as 'quasi sine-wave') inverters are better than square-wave ones, as they provide a less harmonic distortion of the output signal. These inverters are more acceptable for a wide variety of loads – not only lights and small resistive heating appliances but also motors and some electronic equipment, such as TV sets and hi-fi stereos. Due to the still high level of inverter noise, however, modified square-wave inverters are not recommended for other types of electronic devices, clocks, microwave ovens or battery sets for cordless tools.

Sine-wave inverters provide a high-quality waveform. They are suitable for all residential applications due to the low level of noise and harmonic distortion. Sine-wave inverters are a must for every grid-tied solar electric system.

Although the most expensive ones, sine-wave inverters are the best choice, since they are suitable for any applications and comply best with regulatory requirements.

Important!

Have in mind that certain electronic equipment such as mobile phones, microwave ovens, computers, vacuum cleaners, etc. might have problems operating with quasi sine-wave inverters. Furthermore, quasi sine-wave inverters might create additional noise to audio and TV equipment.

Inductive loads, such as fridges, pumps, drills, etc., must be powered by a pure sine-wave inverter.

Furthermore, an inverter must be capable of providing a start-up current for such loads, that is usually 2 to 3 times higher than their nominal operating current.

Square-wave inverters are of worse quality than quasi sine-wave ones.

Important!

The most important features of stand-alone inverters are:

- Generating a stable, sinusoidal AC voltage.
- Providing enough power for all the connected electrical devices.
- Withstanding to electrical surges created by loads with motors.
- Low energy consumption in standby mode.
- Withstanding to tolerances in battery voltage.
- Alerting when battery capacity is low.
- Battery charging – converting the AC voltage coming from a backup generator (if any) to DC voltage for charging the battery bank.
- Overcurrent protection.

With the idea of protection, stand-alone inverters are designed to switch off automatically when the battery voltage falls below a certain level. Therefore, in a battery-based system, the battery charging level should be regularly checked.

Often a stand-alone inverter and a charge controller are combined in a single device. This leads to a lower cost of building the PV system. Such a concept, however, has a notable drawback. Since a couple of devices are integrated into one, the system designer has less freedom, both upon selecting and sizing the individual blocks and components.

Part of the DC power generated by the PV array is used for battery charging, and if the available capacity of the battery is sufficient, the other part of the DC power is turned into AC power to feed the household devices and loads.

When the battery capacity reaches a certain minimum permissible level, the inverter could either alert the system operator either to switch to an additional generator (wind, diesel, etc.) or itself to automatically start such a generator to prevent overdischarging of the battery bank.

Often you could have the following scenario. A stand-alone PV system with an inverter-charger connected to the grid, while the system itself not connected to the grid. If an inverter-charger in a stand-alone solar system is plugged into the grid, such a device consumes AC electricity from the grid to operate and does not send AC electricity to the grid, like grid-direct inverters do. So, you have a stand-alone system and a stand-alone inverter/charger operating by electricity from the grid.

Stand-alone inverters are produced in various power outputs, depending on the type and size of the PV systems. There are 100 W inverters for a small stand-alone system, and there are 5 kW inverters for providing power to all the possible loads in a household.

Certainly, to enable providing more power for fully meeting the energy needs of your home, inverters could be connected by a communication cable.

Another important feature of the battery-based inverters is that their DC input is available just for a limited number of DC voltages (12V, 24V, and 48V), due to the reason that the inverter input appears to be the battery output that comes in exactly those DC voltages.

With grid-direct inverters the situation is different – as you remember, the inverter's input is the PV array's output. The output voltage, however, can vary widely in voltage due to the opportunity for connecting a various number of PV modules into a string.

If you use both the utility grid and a generator as AC power source, you need a grid-direct battery-based inverter supporting multiple power sources.

Important!

Regardless of who is going to install a stand-alone inverter, the following rules are to be kept to ensure maximum safety:

- The inverter should be mounted high enough above the ground to eliminate access to it by children and animals.
- Every inverter should be installed in a dry and well-ventilated place, with enough space available for heat dissipation.
- The inverter, along with its cabling, must be kept away from any flammable materials – gas, oils, solvents and other volatile substances that can easily be ignited by an accidental spark produced by the inverter.

Mounting the inverter in a small room, with not enough ventilation near flammable materials can result in fire and even explosion!

Specifications of stand-alone inverters

- Rated input power – usually selected to be of 20% less than the PV array peak power, due to the various types of losses in solar modules. However, it should be checked whether the inverter's rated input power is higher than the power the solar array is designed to produce at lower ambient temperatures.
- Rated output power – should be enough, so that the inverter would be able to handle all the loads that are to be on simultaneously.
- DC input voltage accepted from the battery bank – voltage values are standard, the most common ones are 12V, 24V, and 48V.
- Output voltage – usually 120 VAC or 240 VAC for most residential buildings.
- Output frequency – 50Hz in Europe, 60 Hz in the USA.
- Surge capacity – allows an inverter to supply much more output power than its rated value within a short period, to provide high starting current to motors (in refrigerators, water pumps, etc.).

Comparison between grid-tied and stand-alone inverters

Important!

Similarities between grid-tied and stand-alone inverters:

- Both types convert DC voltage into AC voltage – 120 VAC or 240 VAC.
- Both types need a source of steady power supply to operate. For grid-direct inverters, such a steady source of power is the grid. For stand-alone inverters, such a steady source is the battery bank.

Differences between grid-tied and stand-alone inverters:

- Grid-direct inverters get their power supply from the grid, while stand-alone inverters are powered by the battery.
- Grid-direct inverters are provided with an anti-islanding feature – they are designed to turn off in case of a grid failure. Stand-alone inverters are not provided with anti-islanding feature, since they are disconnected from the grid.
- Grid-direct inverters are typically high-voltage inverters (up to 700 VDC input voltage), while stand-alone inverters are usually of a low input voltage (12 VDC, 24VDC, and 48 VDC).
- The power input of a grid-direct inverter must be matched to the power input of the solar array. For stand-alone systems, the power output of the PV array does not need to be matched to the inverter power input because in such a case, the

inverter is either connected to the battery or the third terminal of the charge controller.

- For grid-direct inverters, the ground-fault protection (GFP) is a standard feature, while for stand-alone inverters it is not.
- Grid-tied inverters are usually suited for outdoor installation. On the opposite, grid-off inverters are not weatherproof and should be placed indoors, close to the battery bank.

Common mistakes upon connecting the solar system components

Connections are the veins of every solar electric system.

Here are some general rules you must keep:

- AC loads should be connected to the inverter's output while DC loads should be connected to the charge controller's output.
- Certain appliances, such as low-voltage refrigerators, must be connected directly to the battery.
- In a small DC system with a charge controller, you do not need any fuses other than the one incorporated in the charge controller. In larger DC systems, you need to provide a fuse to the positive terminal of the battery.
- A charge controller should always be mounted close to the battery, since precise measurement of the battery voltage is an important part of charge controller's functions. Therefore, even the smallest voltage drops must be avoided.
- A typical charge controller has three terminal connections – for the array, for the battery, and for the DC loads. The charge controller disconnects the battery to prevent it from overcharging and disconnects the DC loads connected to the controller 'DC load' terminal to prevent the battery from overdischarging.
- Every device connected directly to the battery instead of the 'DC load' terminal of the charge controller renders the charge controller battery's overdischarching prevention function useless.
- The inverter should be directly connected to the

143

charge controller 'DC load' terminal.
- When connecting the inverter to the charge controller 'DC load' terminal, check in the charge controller data sheet whether this terminal is powerful enough to provide the input current to the inverter. Otherwise, connect the higher power inverter directly to the battery bank. In such a case, you will render the charge controller's function that prevents the battery from overdischarging useless.

Another option is to use a second charge controller from the same or another manufacturer and set it in overdischarging prevention mode. The second charge controller usually sustains high DC currents and is connected directly to the battery for the purpose of load control. On the other hand, the inverter or DC loads are connected to this second controller.

The third option is to find a battery inverter with regulated low voltage disconnect (LVD) that coincides with the LVD parameters of your battery bank. Such an inverter could be connected directly to the battery bank; however, its overdischarging prevention function would be unreliable.

The fourth option is to a use standalone low-voltage DC disconnect device. This device is connected directly to the battery. Then the DC loads or the inverter are again connected directly to such device. You can find devices that support up to 200A DC currents.

Important:

Cheap charge controllers have a low-current 'DC load' terminal. Their only function is to prevent the battery from overcharging. You can only connect a low-power 12V lamp or other low-power DC device to this terminal. It switches off to prevent the battery from overdischarging.

In such a case, you must connect the rest of the DC loads directly to the battery. There is no way to disconnect the DC loads from the battery in case of overdischarging.

Now, let's review some commonly made mistakes upon connecting the inverter to the solar system.

For example, you use an inverter to get AC voltage (110V-120V/220V-240V) from the DC voltage (12V/24V/48V) produced by a solar panel array connected to a charge controller and a battery bank.

A mistake often made by DIY enthusiasts is connecting the inverter directly to the battery. By doing so, they expose the battery bank to a risk of possible overdischarging.

The inverter should always be connected to the charge controller. In this case, when the battery is running low, to prevent the battery from overdischarging, the charge controller will disconnect the inverter and the load.

There is an exception to this rule – you can connect the inverter directly to the battery bank when you consider the value of the load more important than the battery bank, and you are ready to pay more for more frequent replacement of your batteries, by regularly and intentionally shortening their lifecycle.

Such a situation could happen, for example, when the load is a fridge containing a life-saving medicine or an important radio device.

If the inverter input DC current is higher than the charge controller DC load rated current, you should connect the inverter directly to the battery bank. In such a case, you should have in mind the following.

Important:

When connecting inverter directly to the battery, you can't rely on the inverter's low voltage shut-off function to prevent a battery from overdischarging because its main function is to protect the inverter, not the battery.

Inverter's low voltage shut-off value is too low to prevent a battery from overdischarging.

There is a strict sequence to follow upon introducing the charge controller to the solar electric system while connecting or disconnecting the wires between the solar panel, charge controller, and battery bank:

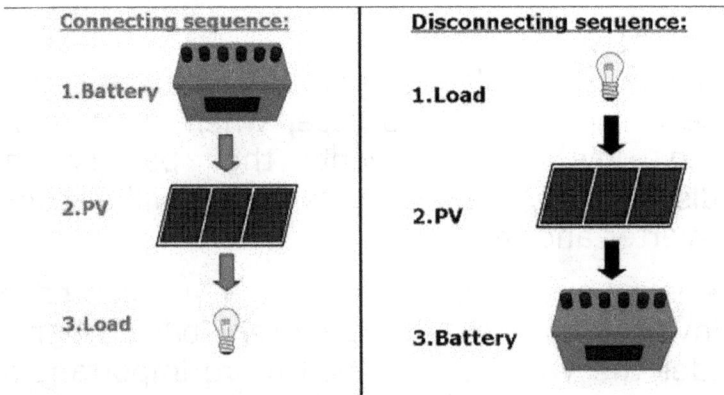

If the battery is not connected to the charge controller **first**, higher solar panel voltage could damage the load!

As you know, the charge controller connects or disconnects the load on the basis of the voltage related to the battery's state of charge.

For example, the solar panel voltage of an array in a 12V system may reach up to 18-19V, while the maximum input voltage of a 12V-rated device is about 14V.

Also by the authors:

The Truth About Solar Panels: The Book That Solar Manufacturers, Vendors, Installers And DIY Scammers Don't Want You To Read
[Paperback and all types of eReaders editions – Kindle, Kobo, Nook, Apple, etc.]
ISBN-10: 6197258013
ISBN-13: 978-6197258011

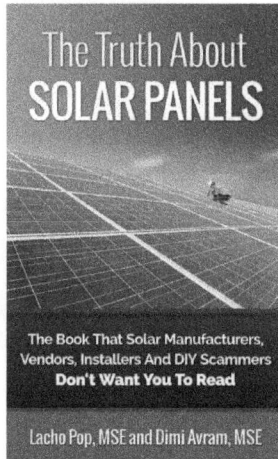

The book is about solar photovoltaic panels – the main building unit of solar systems. By reading it you get a complete know-how to buy, build, compare, evaluate, mix and assemble different type of solar panels.

It describes the basic solar panel types and reviews in details their main features and parameters. The book makes a unique presentation of cheap and second-hand solar panels by describing their pros and cons, where to find them, how to assess them and how to use them. Special attention is paid to DIY solar panels by providing a proven methodology of evaluating a small solar electric system based on DIY panels. All the information is presented in an easy-to-read manner, by lots of examples, recommendations and tips, and proper summaries where needed.

The book is targeted to various groups of readers – homeowners,

do-it-yourself solar enthusiasts, lovers of recreational vehicles, campers, boats and other outdoor activities, survivalists, potential investors in solar power, business owners interested in solar power, students, teachers, people interested in becoming solar installers, people working in the sales and marketing area of solar power and green energy industry and many others keen in solar power and renewable energy.

The Ultimate Solar Power Design Guide: Less Theory More Practice (The Missing Guide For Proven Simple Fast Sizing Of Solar Electricity Systems For Your Home, Vehicle, Boat or Business)
[Paperback and all types of eReaders editions – Kindle, Kobo, Nook, Apple, etc.]

ISBN-10: 6197258048
ISBN-13: 978-6197258042

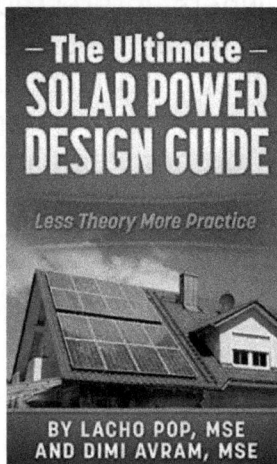

The Ultimate Solar Power Design Guide is a straightforward guide on solar power system sizing. It is written by experts for beginners and professionals alike.

Proper sizing of a solar panel system is crucial. The final goal of sizing a photovoltaic system is to come up with a cost-effective, efficient, and reliable photovoltaic system for your home, RV vehicle, boat or business – a solar panel system that squeezes out the maximum possible power for every cent invested.

The main drawback to the majority of solar books is that they provide too much general information about solar panels and solar components and, if you are lucky enough, just one or two very basic sizing formulas.

The mission of this book is to fill this gap by offering a simple, practical, fast, step-by-step approach to sizing a solar panel system of any scale, whether simple or complex, intended for your home, business, RV vehicle or boat.

The book is written by experts holding master's degrees in Electronics, and it is targeted for those who cannot get started or are utterly confused.

It contains lots of fast and simple universal sizing methods applicable to all cases, accompanied by proper sizing examples. Thanks to this approach, you will be capable of sizing any solar power system or tailor the sizing methods according to your needs.

Top 40 Costly Mistakes Solar Newbies Make: Your Smart Guide to Solar Powered Home and Business 2016 Edition

[Kindle and Paperback Edition], Kindle ASIN: B01GGB7QP8, Paperback ISBN-13: 978-6197258073

ISBN-10: 6197258072
ISBN-13: 978-6197258073

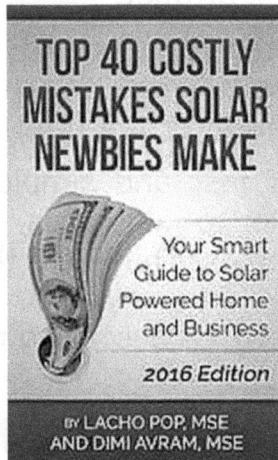

The book "Top 40 Costly Mistakes Solar Newbies Make" is a simple and practical guide that could save you a lot of money, headaches and time during the planning, buying, implementation, and operation phase of your solar power system.

Whether you have decided to buy a solar electricity system or assemble it yourself, you need a simple and easy-to-follow step by step guide.

There are thousands of books, articles, leaflets, forums and many other resources available online, written by unqualified authors, telling you what to do.

They could be misleading you!

By reading this book you get a trustworthy information

in the form of practical tips and hints given by engineers with a long-term experience in electrical and electronic engineering.

Here are the most common groups of mistakes commonly committed by people who have decided to go solar:

- General mistakes and misconceptions
- Mistakes during location assessment
- Mistakes with solar panels
- Mistakes in solar system sizing
- Mistakes in assembling the system components
- Mistakes in buying a solar system

Mistakes in solar power system maintenance. Although solar electrics are everywhere around us and there are tons of resources available on this topic, you don't need to be an expert to assess what is good for your specific case.

Neither you have to refer to costly solar consultants to reveal you the solar secrets.

What you need is this simple and easy-to-read solar book. It can save you both money and trouble down the road. It is frankly written by experts for everyone who cannot get started.

Get started your solar journey today!

Glossary of Terms

Alternating current (AC) – electrical current changing its direction at a given interval.

Balance of system (BoS) equipment – all the equipment apart from the solar array, which is needed for a solar electric system to operate.

Battery – a device capable of producing DC electricity and storing it for later use.

Battery bank – a combination of batteries connected together.

Cable Losses – the overall system losses due to cable resistance.

Capacity of a battery – the amount of electricity a battery can store. Capacity is measured in Ampere-hours (Ah).

Charge controller – a device managing the process of battery charge and discharge.

Current – directional movement of electrons upon some voltage applied.

Conductor – stuff where electric current can occur.

Days of Autonomy (DoA) – the desired number of consecutive days that we would like the battery bank to power the load in case of a complete lack of sunshine.

Depth of Discharge (DoD) – defines up to how much percentage the battery bank should be discharged: 100%=empty battery bank; 0%= full battery bank

Disconnect (breaker) – an electric switch protecting an electric circuit from overload.

Direct current (DC) – an electric current always flowing in the same direction.

Distribution panel (distribution board) – a device dividing electrical power supply into several electrical circuits.

Energy – the work that can be done within a period.

Energy efficiency – a set of measures resulting in electrical consumption reductions.

Fuel generator – a generator working on combustive fuel, able to generate AC electricity.

Grid-tied (grid-direct, grid-connected, grid-on) system – a solar electrical system producing electricity that can be both used in your home/office and exported to the grid.

Hybrid system – an off-grid system that combines photovoltaics with additional power sources (e.g., combustive fuel generator, wind generator, etc.) to produce electricity.

Insulator – stuff in which no electric current can flow.

Inverter – a device converting DC into AC electricity.

Load – a device consuming electricity to do some useful work.

Net metering – the process of measuring the solar electricity exported to the grid by a solar panel system owner, credited by the local utility company.

Off-grid (autonomous) system – a solar electrical system disconnected from the grid and producing electricity for home/office use only.

Photovoltaic (solar) cell – the smallest

semiconductor unit producing electricity when exposed to sunlight.

Photovoltaic (solar) generator – solar array with all its cabling and disconnects

Photovoltaic (solar) module – a combination of solar cells connected together.

Photovoltaic (solar) panel – a combination of solar modules connected together. Often, although not fully correct, terms 'solar panel' and 'solar module' are used interchangeably.

Photovoltaic (solar) array – a combination of solar panels connected together.

Photovoltaic (solar) system – a combination of solar modules and other equipment connected to produce electricity for practical needs.

Power – the rate of consuming/generating energy.

Semiconductor – stuff where electric current can only occur under certain conditions. Solar panels are made of semiconductor (silicon) material.

Stand-alone system – an off-grid system that only relies on photovoltaics to produce electricity.

Voltage – a difference in the potential (hidden) energy between two points, causing current to flow upon free electrons available.

References

1. Antony, Falk, Christian Durschner, Karl-Heinz Remmers. 2007. Photovoltaics for Professionals: Solar Electric Systems Marketing, Design and Installation, Routledge.
2. Boxwell, Michael. 2012. Solar Electricity Handbook, Greenstream Publishing, Amazon Kindle Edition
3. Brooks, Bill. Designing and Installing Code-Compliant PV Systems – from materials developed by Endecon Engineering and the Florida Solar Energy Center.
4. Clean Energy Council, Australia. 2002. Grid-Connected PV Systems – System Design Guidelines for Accredited Designers, Issue 3 July 2007, November 2009 Update.
5. Clean Energy Council, Australia. 2008. Electricity from the Sun – Solar PV systems explained, 3rd Edition, June 2008
6. Endecon Engineering. 2008. A Guide to Photovoltaic (PV) System Design and Installation, June 14, 2001.
7. German Energy Society. 2008. Planning and Installing Photovoltaic Systems – a Guide for Installer, Designers and Engineers.
8. Mayfield, Ryan. 2010. Photovoltaic Design and Installation for Dummies, Wiley Publishing Inc.
9. Munro, Khanti. 2010. Designing a Stand-Alone PV System, Home Power Magazine 136, April-May 2010, pp.78-84.
10. Pop MSE, Lacho, Dimi Avram MSE. (2015-10-26).The Truth About Solar Panels: The Book That Solar Manufacturers, Vendors, Installers And DIY Scammers Don't Want You To Read, Kindle

Edition. Digital Publishing Ltd.

11.   Pop MSE, Lacho, Dimi Avram MSE. 2015. The Ultimate Solar Power Design Guide: Less Theory More Practice, Kindle Edition. Digital Publishing Ltd.

12.   Pop MSE, Lacho, Dimi Avram MSE,(2016-07-27),Top 40 Costly Mistakes Solar Newbies Make: Your Smart Guide to Solar Powered Home and Business, Paperback Edition, Digital Publishing Ltd.

13.   Sandia National Laboratories. 1991. Maintenance and Operation of Stand-Alone Photovoltaic Systems.

14.   SEAI (Sustainable Energy Authority of Ireland), Best Practice Guide Photovoltaics (PV).

15.   Solar Energy International. 2007. Photovoltaics: Design & Installation Manual, New Society Publishers.

Web Resources:

http://www.usa-eds.com
http://openclipart.org
http://wikipedia.org

Index

www.ingramcontent.com/pod-product-compliance
Lightning Source LLC
Chambersburg PA
CBHW050117210326
41519CB00015BA/3997